Atmospheric Aerosols: Physics and Chemistry

Atmospheric Aerosols: Physics and Chemistry

Edited by **Luke McCoy**

New York

Published by Callisto Reference,
106 Park Avenue, Suite 200,
New York, NY 10016, USA
www.callistoreference.com

Atmospheric Aerosols: Physics and Chemistry
Edited by Luke McCoy

International Standard Book Number: 978-1-63239-082-0 (Hardback)

Contents

Preface

The purpose of the book is to provide a glimpse into the dynamics and to present opinions and studies of some of the scientists engaged in the development of new ideas in the field from very different standpoints. This book will prove useful to students and researchers owing to its high content quality.

This book discusses the characteristics of atmospheric aerosols and their effect on climatic conditions. Ground-based, air-borne and satellite data have been collected and analyzed. The measurement techniques and atmospheric conditions are provided in a detailed account. The book also includes extensive information on the organic and inorganic elements of atmospheric aerosol. The author explicates how aerosol particles are chemically and physically processed, temporally and spatially distributed, emitted, formed and transported. This book caters to the need of researchers to resolve several issues related to the complex interaction between atmospheric aerosols and climatology.

At the end, I would like to appreciate all the efforts made by the authors in completing their chapters professionally. I express my deepest gratitude to all of them for contributing to this book by sharing their valuable works. A special thanks to my family and friends for their constant support in this journey.

Editor

Aerosols Chemistry and Physics

Review of Aerosol Observations by Lidar and Chemical Analysis in the State of São Paulo, Brazil

Gerhard Held, Andrew G. Allen, Fabio J.S. Lopes,
Ana Maria Gomes, Arnaldo A. Cardoso, Eduardo Landulfo

Additional information is available at the end of the chapter

1. Introduction

Large-scale forest fires in the tropics, emitting vast amounts of aerosols and trace gases, drew the attention of scientists around the world in the late 80s and early 90s. A number of international collaborative research projects, such as TRACE-A (Transport and Atmospheric Chemistry near the Equator-Atlantic, [1]) and SAFARI-92 (South African Fire-Atmosphere Research Initiative, [2]), were initiated under the auspices of the International Geosphere-Biosphere Programme to investigate biomass burning emissions and their long-range transport. One of the areas of great interest was the Amazon region (Figure 1), which later led to the creation of the international Large Scale Biosphere-Atmosphere Experiment in Amazonia (LBA, [3]) in Brazil in 1998. Several intense observation campaigns were dedicated, not only to rainfall measurements by radar and storm structure, but also to biomass burning, monitoring of emissions and transport of aerosols and their impact on the vegetation and population of the region. However, monitoring of background concentrations of aerosols, deploying stacked filter units, had already been initiated in 1990 at the "Sierra do Navio" site (Amapá, about 190 km north of the equator) and in Cuiabá (Mato Grosso), a town located in the Brazilian savannah [4]. The location of both sites is shown in Figure 1.

São Paulo is Brazil's most populous State, with approximately 42 million inhabitants (21,5% of Brazil's total population) in an area of 249 000 km². The region is diverse in terms of its geography, natural environment and economy, and can be broadly classified into three main zones. In the southeast, the Atlantic coastal strip is separated from the remainder of the State by the scarp of the Serra do Mar, containing Brazil's largest remaining areas of

Atlantic rainforest, a threatened ecosystem that has been largely eliminated in most of the Brazilian States bordering the Atlantic ocean. Located on a plateau above the scarp are the densely populated and heavily industrialized regions of metropolitan São Paulo (RMSP) and its satellite cities. Continuing inland, the largest fraction of the area of the State has an economy mostly based on agro industry. Here has been widespread conversion of natural ecosystems to agriculture. The most important single agricultural activity is sugar cane production, although there are also substantial cattle ranching, citrus cultivation and agro forestry for pulping and construction. In all regions, it is largely local emission sources that determine the chemical composition of the atmospheric aerosol, with a smaller influence of long-range transport of polluted air masses from elsewhere in Brazil.

Figure 1. Brazil, showing the location of São Paulo State in relation to the Amazon region, as well as the background monitoring stations in Sierra do Navio (Amapá) and Cuiabá (Mato Grosso).

In terms of atmospheric quality, suspended aerosol particles are (together with ozone) probably the most important atmospheric pollutant in both São Paulo city and the largely agricultural hinterland of the State. Ozone is generated during reactions involving the nitrogen oxides (NO_x) and volatile organic compounds (VOCs) emitted from vehicles, biomass burning and biogenic sources. The particulates are either emitted directly (in the form of primary aerosols), or are produced during reactions involving gaseous precursors (SO_2, NO_x and hydrocarbons). In large urban areas, such as the Metropolitan Region of São Paulo (RMSP), anthropogenic emissions from vehicles and industrial processes are the dominant contributors to elevated aerosol levels, while biomass burning [5-7] and dust lifted from barren fields (Figure 2) during the dry winter season constitute the principal sources of aerosols in the central and western sectors of the state. The State of São Paulo is the largest producer of sugar cane in Brazil, accounting for about 60% of Brazil's harvest [8], with more than 4,7 million hectares planted in 2010, of which 44% are burnt before harvesting [9]. The sugar cane is mostly harvested from April to November. Although progress is being made in mechanization, large areas are still harvested manually, which requires burning of the crop in sectors of the plantations during the night prior to manual cutting to remove excess foliage. This practice results in large quantities of aerosols and trace gases being emitted into the atmosphere (Figure 2a), not only negatively affecting local towns, but also regions much further downwind [10-12], demonstrating the importance of monitoring aerosols throughout the State.

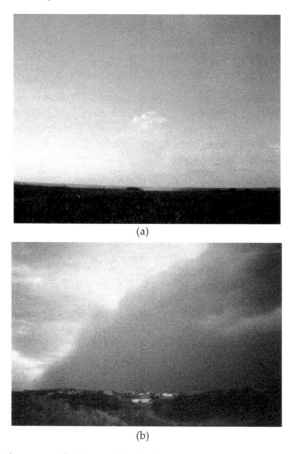

(a)

(b)

Figure 2. (a) Typical sugar cane fire in central São Paulo State. (b) Dust lifted from freshly cut sugar cane fields by the downdraft of an approaching storm.

Along the São Paulo coast, marine aerosols are modified by the inclusion of pollutants emitted from transport, urban, and industrial sources. There are some areas where levels of anthropogenic pollution are low, and where the aerosol composition can be mainly attributed to natural origins. However, compared to metropolitan São Paulo and the interior of the State, the coastal zone has been much less well studied, with the exception of Cubatão, a heavily industrialized town near the coast close to Santos.

In the State of São Paulo, the first aerosol measurements began in Cubatão [13], and within the metropolitan area of São Paulo, notorious for its traffic emissions [14]. In terms of morphology, São Paulo is among the world's five largest cities, and is sixth largest in terms of population [15], with about 11,3 million inhabitants. The population of the Metropolitan Region of São Paulo (RMSP), which includes peripheral urban areas, reached an estimated 19,9 million persons in 2009 [16]. Human activities including road transport and industry now exert an enormous impact on air quality in the region, and therefore on the health of the population [17]. The total fleet of vehicles (cars, buses, trucks and motorcycles, powered by gasoline, ethanol and diesel) in the State of São Paulo exceeded 12,8 million in 2011, of which about 50% operate within the RMSP [9].

Observations from the Brazilian Lightning Detection Networks (RINDAT [18] and BrasilDAT at ELAT/INPE [19]) have shown a significantly higher lightning frequency over the RMSP and other large urban complexes within the State since the inception of the RINDAT Lightning Network in 1999 [20, 21]. This prompted a study of the impact of anthropogenic emissions on the frequency of lightning [22], showing a distinct increase of cloud-to-ground flashes, not only over the RMSP, but also over other large cities and densely populated or industrialized regions in the State, correlated to the occurrence of heat islands and increased concentrations of PM_{10}.

2. Meteorology and climatology of the State of São Paulo

Since the meteorology of a region has a major impact on the dispersion or accumulation of pollutants, a brief characterization of the climate is appropriate. The State of São Paulo is located between the latitudes of about 20° and 25° South (Figure 1), thus falling into the transition zone from a tropical to a subtropical climate, with an annual rainfall total ranging between 1250 and 1650 mm in the interior, increasing to 1850 mm over the narrow coastal strip [23]. The year can be roughly divided into two periods, *viz.*, the rainy season from October to March, when most of the rain is produced by convective storms, and the dry winter months from April to September. During the rainy season, conditions are more representative of the tropical climate, with the occasional occurrence of a South Atlantic Convergence Zone (SACZ), which can be identified from satellite images as a cloud band with orientation northwest to southeast, extending from the southern region of Amazônia into the central region of the South Atlantic Ocean [24]. The SACZ situations can last more or less continuously from 4 days to more than one month and are extremely efficient producers of rain in the form of tropical thunderstorms, with accompanying high humidity. During the relatively dry winter months, the climatic conditions are more typical of the subtropics, with only occasional heavy rainfalls being caused by the passage of baroclinic systems (mostly cold fronts), moving from southwest to northeast across the State, but for the remaining time, the weather is dominated by a high pressure system, resulting in elevated temperatures, with low humidity and high stability in the Planetary Boundary Layer, favoring the accumulation of pollutants in the atmosphere of the region [25].

Sodar observations made during the period of June 2009 to December 2011 showed that strong nocturnal Low-Level-Jets (LLJs) develop on top of the surface radiation inversion, mostly during the relatively dry austral winter months (May – October), when stable conditions prevail [26, 27]. These LLJs generally form during the late evening at altitudes ranging from 250–500 m AGL, with maximum speeds of 12–20 m.s^{-1}. They usually last until 08:00–09:00 Local Time (LT), when the inversion has been eroded by the solar radiation. The frequency of LLJs varied from 3 - 22 days per month, with higher frequencies and greater intensity generally during the winter months. Observations with a sodar were made at three different locations in the central region of the State, *viz.* in Bauru, Rio Claro and Ourinhos. Earlier measurements, deploying tethered balloons and radiosondes in the eastern region of the State, yielded similar results in terms of structure, dynamics, seasonality and development characteristics [28]. LLJs have been observed in many parts of the world and were found to have regional extent. The practical importance of the LLJ lies in the rapid transport of moisture and pollutants in a narrow vertical band above the radiation inversion [29].

3. Ground-level monitoring of particulates

Regular monitoring of air pollutants under the auspices of the Companhia de Tecnologia de Saneamento Ambiental (CETESB), the air quality "watchdog" in the State of São Paulo, started in the 70s, but a fully automatic monitoring network was only installed in 2000. Since then, observations are available in real time [30]. In 2001, 29 automatic stations, the majority in the RMSP, were already in operation [31]. From 2008 onwards, the automatic monitoring network was significantly expanded. In 2011, 42 monitoring stations in 28 towns were in operation, 19 in the RMSP and 23 in the remaining parts of the State [9]. The majority of the stations monitor particulate matter (PM$_{10}$), NO, NO$_2$, NO$_x$ and O$_3$, as well as meteorological parameters, while a few also measure PM$_{2.5}$, SO$_2$ and CO. The automatic air quality monitoring network is shown in Figure 3. Additionally, CETESB also maintained a network of 41 manual monitoring stations during 2011, where measurements are made of PM$_{2.5}$, PM$_{10}$, TSP (Total Suspended Particulates), black smoke and SO$_2$, in various combinations [9]. Aerosol mass concentrations are determined using either β-attenuation instruments (automatic stations) or gravimetric and reflectometric techniques (manual stations).

In accordance with recommendations of the World Health Organization [32], CETESB defines 5 levels of air quality: "Boa" (good), "Regular" (regular), "Inadequada" (insufficient), "Má" (bad) and "Péssimo" (extremely bad), the highest being invoked if one of the monitored pollutants exceeds the pre-defined threshold. The national air quality standards are defined in CONAMA Resolution No. 03/90 (Table 2 in [9]).

PM$_{10}$ and TSP measurements are available since 1984 and 1985, respectively [31], although initially only from very few stations in the interior of the State, but gradually increasing to 41 and 11, respectively, in 2011 [9]. Figure 4 shows the year-to-year variation of annual mean PM$_{10}$ concentrations against the National Air Quality Standard (PQAr) for the RMSP and two sites in Cubatão (Figure 3, Nos. 24 and 25), which is one of the major industrial hubs in Brazil, where one site is located within the industrial suburb (No. 25) and the other in the town centre (No. 24). A significant reduction of mean annual PM_{10} concentrations can be noticed from 1998 onwards, confirming the success of implementation of stringent air quality control measures, administered by CETESB. However, within the industrial suburb, confined in a valley, concentrations are still about twice the PQAr. A detailed description of Cubatão, its industrial activities and their location are found in [33]. More details on current PM_{10} and TSP concentrations are provided in Section 4.3.

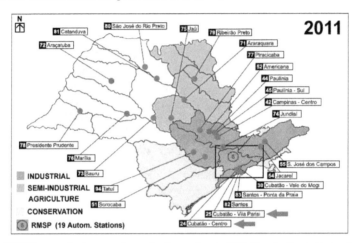

Figure 3. CETESB network of automatic monitoring stations in 2011. The shading indicates the principal land use in four schematic regions of the State, directly related to the type of emissions. Adapted from [9].

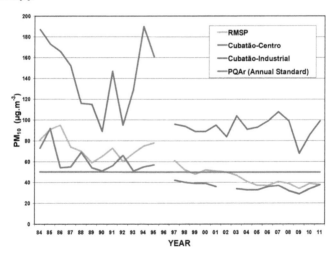

Figure 4. Year to year variation of PM_{10} from 1984 – 2011 for the RMSP and Cubatão. The data were extracted from [9, 31]. PQAr = 50 $\mu g.m^{-3}$ represents the Annual Standard.

Figure 5 demonstrates that the air quality in the interior of the State from 2002 to 2011, when a reasonable number of monitoring sites were already in operation [9], was generally well below the annual standard of 50 µg.m^{-3} for PM$_{10}$, with the exception of Santa Gertrudes, just south of Rio Claro (Figure 6), where several large ceramic industries are located, notorious for emitting large quantities of aerosols. At two other monitoring sites, annual means were close to the Annual Standard. At Limeira mixed industrial activities range from metallurgical, through cellulose to ceramics, besides sugar cane and orange production and processing plants. Limeira and Santa Gertrudes are medium-sized industrial towns, about 20 and 40 km northwest of Americana (Figure 3, No. 52). The other site is in Piracicaba (Figure 3, No. 77), which also hosts mixed industrial activities, including a significant petrochemical plant. However, the exceedance in 2011 was most likely caused by major road construction works in the immediate vicinity of the monitoring site [9].

Although annual mean concentrations of PM in the State of São Paulo seem to be quite acceptable, it is obvious that violations of the daily Air Quality Standard do occur occasionally in several towns of the interior and within the RMSP. Comprehensive annual and specialized technical reports and publications on the air quality in the State of São Paulo, including detailed monitoring results, are available online [9].

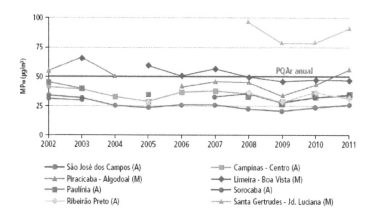

Figure 5. Year to year variation of PM$_{10}$ from 2002 -2011 for monitoring sites in the interior of the State of São Paulo (after [9]). PQAr = 50 µg.m^{-3} represents the Annual Standard.

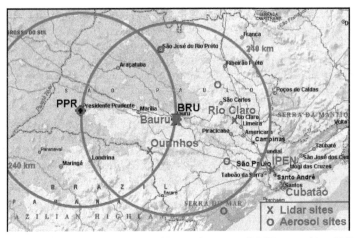

Figure 6. Aerosol monitoring sites in the State of São Paulo (except CETESB network) and 240 km ranges of IPMet's radars in Presidente Prudente (PPR) and Bauru (BRU). Sites from where lidar measurements are available are marked with x.

4. Chemical composition of aerosols

4.1. Agro-industrial rural regions

Seasonal variability in the major soluble ion composition of atmospheric particulate matter in the principal sugar cane growing region of central São Paulo State indicates that pre-harvest burning of sugar cane plants is an important influence on the regional-scale aerosol chemistry [34]. The size-distributed composition of ambient aerosols is used to explore seasonal differences in particle chemistry, and to show that dry deposition fluxes of soluble species, including important plant nutrients, increase during periods of biomass (sugar cane trash) burning [6, 10].

Concentrations of trace gases and aerosols were determined at six measurement sites of a regional network in São Paulo State (blue circles in Figure 6), installed in rural areas including the State's central agricultural zone and the eastern coast [11] as part of an experimental research project to determine the anthropogenic component of nutrient deposition. The measurements were made over 12 months during 2008/2009 (one week of continuous sampling per month). Aerosols were collected onto 47 mm diameter Teflon filters using active samplers, and trace gases (NO_2, NH_3, HNO_3 and SO_2) were sampled using diffusion-based devices. The soluble ions NO_3^-, NH_4^+, PO_4^{3-}, SO_4^{2-}, Cl^-, K^+, Na^+, Mg^{2+} and Ca^{2+} were analyzed in aqueous extracts of the aerosol filters, using ion chromatography. NO_2, HNO_3 and SO_2 were similarly determined as NO_2^-, NO_3^- and SO_4^{2-}, following aqueous extraction of the collection media. NH_3 was determined using a colorimetric technique. Identification and quantification of nutrient sources was achieved using principal component analysis (PCA) followed by multiple linear regression analysis (MLRA) applied to the chemical data. Dry deposition fluxes were estimated using the measured atmospheric concentrations together with dry deposition velocities of gases and aerosols to different surface types, including tropical forest, savannah, sugar cane, pine, eucalyptus, orange, coffee, pasture and water. The annual cycle in deposition, to a sugar cane surface, of reactive nitrogen and sulphur in the gaseous, aerosol and dissolved phases is illustrated in Figure 7.

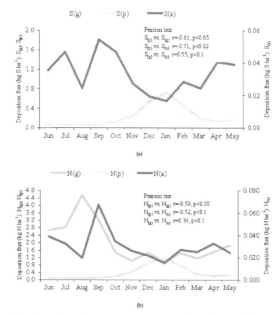

Figure 7. Annual cycle in deposition fluxes to a sugar cane surface of: **(a)** sulphur in gaseous ($S_{(g)}$), aerosol ($S_{(a)}$) and rainwater ($S_{(p)}$) phases; **(b)** nitrogen in gaseous ($N_{(g)}$), aerosol ($N_{(a)}$) and rainwater ($N_{(p)}$) phases. Primary y-axes: gas and rainwater; secondary y-axes: aerosol. Data for Araraquara.

The sugar cane industry has a major impact on air quality and the characteristics of the atmospheric aerosol. During the dry season (May to October), the burning of the cane, a prerequisite of manual harvesting, has for many years resulted in very large emissions of pollutants, including high carbon content aerosols. These particles contain water-soluble organic carbon (WSOC), anions (sulphates, nitrates and chlorides), cations (potassium, ammonium, calcium, magnesium, sodium), black carbon (BC), insoluble organic carbon and trace metals. Carbonaceous material comprises the bulk of the aerosol mass, especially in fine particles [5-7, 35-39]. In 2004, the annual emission of nitrogen oxides (NO_x) from sugar cane burning in Sao Paulo State was in excess of 45 Gg.N [40]. This is not only indicative of the scale of the emissions, but also of their potential for formation of secondary aerosols (containing nitrates, amongst other components).

In 2011, annual mean PM_{10} concentrations measured at automatic monitoring stations in the agro-industrial interior of São Paulo State were in the range 23-91 $\mu g.m^{-3}$, with the highest values at locations affected by primary emissions from ceramics industries (Figure 5). At sites in the sugar cane production areas, annual mean PM_{10} concentrations were in the range of 32-41 $\mu g.m^{-3}$ [9]. The data revealed no obvious trends in PM_{10} concentrations during the period 2002-2011 (Figure 5).

A proportion of the primary material emitted during sugar cane fires is in the form of the large ash fragments notorious for causing domestic soiling problems in the region. During 1995-1996, CETESB investigated deposition rates, and measured the concentrations of PAHs, PCBs, dioxins and furans. The sedimented material was collected during the harvest period using plastic funnels lined with polyurethane foam, positioned near to plantations and in the urban area of the city of Araraquara. Samples were also collected in parallel using a high volume filter-based sampler. Levels of PCBs were in the range 4-12 $ng.m^{-3}$, and showed no association with levels of carbonaceous material derived from the fires. Deposition fluxes of the dioxins and furans were in the range 1-17 $pg.m^{-2}.day^{-1}$, and were higher in greater proximity to plantations, indicating that sugar cane burning was a source of these compounds. The PAHs were found in two distinct groups. Naphthalene, fluorene, phenanthrene, anthracene, fluoranthene, and pyrene were present at concentrations exceeding 30 $ng.m^{-3}$, while acenaphthene, chrysene, benzo(a)fluoranthene, benzo(k)fluoranthene, benzo(a)pyrene, dibenzo(a,h)anthracene, benzo(g,h,i)perylene and indeno(1,2,3,c,d)pyrene were found at up to 21 $ng.m^{-3}$. Concentrations were always higher during the harvest period [41].

The presence of PAHs in ash from sugar cane fires was also reported by Zamperlini et al. [42, 43]. In the PM_{10} fraction, it was found that the most abundant polycyclic aromatic hydrocarbons were phenanthrene and fluoranthrene, and the least abundant was anthracene [44]. Cluster analysis of the total PAH concentrations for each day of sampling, and the corresponding meteorological data, suggested that concentrations of PAHs were independent of climatological conditions or season of the year. Vehicular sources were identified during both dry and wet seasons, although sugar cane burning emissions were the dominant source during the dry season.

Sugar cane burning is a major source of acidic gases that contribute to the formation of secondary aerosols. In Araraquara, Da Rocha et al. [36] reported concentrations of 9,0 ppb (HCOOH), 1,3 ppb (CH_3COOH), 4,9 ppb (SO_2), 0,3 ppb (HCl) and 0,5 ppb (HNO_3). Extremely high concentrations of these gases were measured in the plumes downwind of sugar cane fires: 1160-4230 ppb (HCOOH); 360-1750 ppb (CH_3COOH); 10-630 ppb (SO_2); 4-210 ppb (HCl); and 14-90 ppb (HNO_3). Highest levels of SO_2, HCl and HNO_3 in Araraquara were measured during the harvest period, with peak concentrations in the evening (the time of the fires).

The distribution of soluble ionic material between fine (<3,5 μm) and coarse (>3,5 μm) aerosol fractions was determined by Allen et al. [5], who measured the ions $HCOO^-$, CH_3COO^-, $C_2O_4^{2-}$, SO_4^{2-}, NO_3^-, Cl^-, Na^+, K^+, NH_4^+, Mg^{2+} and Ca^{2+}. The fine and coarse particles showed acidic and basic properties, respectively, and concentrations of all major ions increased significantly during the dry season (Figure 8). Da Rocha et al. [6] collected aerosols in twelve size fractions, and used calculation of ion equivalent balances to show that during burning periods, the smaller particles (Aitken and accumulation modes) were more acidic, containing higher concentrations of SO_4^{2-}, $C_2O_4^{2-}$, NO_3^-, $HCOO^-$, CH_3COO^- and Cl^-, but insufficient NH_4^+ and K^+ to achieve neutrality. Larger particles showed an anion deficit due to the presence of unmeasured ions, and comprised re-suspended dusts modified by accumulation of nitrate, chloride and organic anions. Increases of re-suspended particles during the burning season were attributed to release of earlier deposits from the surfaces of burning vegetation, as well as increased vehicle movement on unsealed roads. During the winter months, the relative contribution of combined emissions from road transport and industry diminished due to increased emissions from biomass combustion and other activities specifically associated with the harvest period.

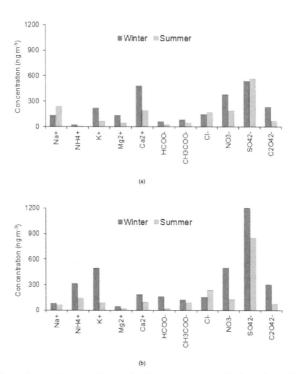

Figure 8. Comparison of aerosol composition in the Araraquara region during winter (biomass burning) and summer (non-burning) periods: **(a)** Coarse particles; **(b)** fine particles.

In separate work, biomass-burning aerosols were found to contribute around 60 and 25% of the mass of fine and coarse aerosols, respectively, in the Piracicaba sugar cane growing region [7]. A high proportion of the elements K, S, Cl, Br, Fe and Si in aerosols has been linked to biomass burning [45], indicative of both a combustion component (emissions of K, S, Cl and Br) and a suspended soil dust component (emissions of Fe and Si).

In a study reported in [38], elemental analysis of individual and bulk aerosols collected in rural areas was followed by evaluation of the data using statistical hierarchical clustering, which revealed the contributions of two different types of carbonaceous material (biogenic and carbon-rich) and two aluminosilicate fractions (pure or mixed with carbon). These findings contrasted with the findings of similar work in the atmosphere of São Paulo city, where hierarchical clustering analysis revealed the presence of metal compounds, silicon-rich particles, sulphates, carbonates, chlorides, organics and biogenic particles [46]. This reflects the very different characteristics of the aerosols found in the two regions.

Da Rocha *et al.* [6] showed that dry deposition fluxes of important plant nutrients increased during the sugar cane burning season. During this period, the fine fraction aerosol was more acidic and contained elevated concentrations of SO_4^{2-}, $C_2O_4^{2-}$, NO_3^-, $HCOO^-$, CH_3COO^- and Cl^-, but insufficient NH_4^+ and K^+ to achieve neutrality. Larger particles consisted of re-suspended dust, modified by inclusion of nitrate, chloride and organic anions. The increases in annual particulate dry deposition fluxes due to higher fluxes during the sugar cane harvest were 44,3% (NH_4^+), 42,1% (K^+), 31,8% (Mg^{2+}), 30,4% ($HCOO^-$), 12,8% (Cl^-), 6,6% (CH_3COO^-), 5,2% (Ca^{2+}), 3,8% (SO_4^{2-}) and 2,3% (NO_3^-). The contributions of dry deposition to total deposition (including precipitation scavenging, excluding gaseous dry deposition) were 31% (Na^+), 8% (NH_4^+), 26% (K^+), 63% (Mg^{2+}), 66% (Ca^{2+}), 32% (Cl^-), 33% (NO_3^-) and 36% (SO_4^{2-}).

Deposition rates of aerosol nutrient species to a range of natural and agricultural surfaces were reported in [10], using a size-segregated particle dry deposition model. Fluxes greatly exceeded those expected under pristine conditions, with deposition to tropical forest found to have increased by factors of 12,2 (NO_3^-), 6,2 (PO_4^{3-}) and 2,6 (K^+) (Figure 9). Source apportionment using principal component analysis (PCA) and multiple linear regression analysis (MLRA) revealed that in central São Paulo State, biomass burning, products of secondary reactions and soil dust re-suspension contributed 43%, 31% and 21% of $PM_{2.5}$ mass, respectively. Re-suspension and biomass burning contributed 22% and 19%, respectively, to PM_{10} mass, and re-suspension accounted for approximately half the mass of coarse particles. At least 40% of NO_3^--N, 20% of phosphorus and 55% of potassium deposited originated from agriculture-related emissions.

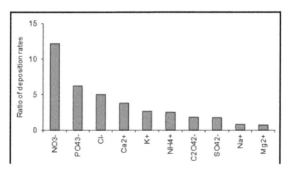

Figure 9. Graph showing the present-day increase in aerosol dry deposition rates to a tropical forest surface, compared to deposition rates estimated for pristine conditions.

Emissions of reactive nitrogen compounds are of concern due to their influence on both atmospheric acidity (production of HNO_3 from reactions involving NO_2) and the formation of photochemical oxidants such as ozone and peroxyacetyl nitrate (PAN). Reactions of acidic species with ammonia generate ammonium sulphates and nitrates, mainly in the long-lived accumulation mode size fraction. Deposition of reactive nitrogen can cause eutrophication of water bodies, as well as the release of trace metals in soils. Machado et al. [47] found that emissions of reactive nitrogen during sugar cane burning, in the forms of NH_3, NO_x and particulate nitrate and ammonium, were equivalent to 35% of the annual fertilizer-N application. The concentrations of nitrogen oxides showed a positive association with the number of fires, reflecting the importance of biomass burning as a major emission source, and mean concentrations of NO_x in the dry season were twice those in the wet season. During the dry season, biomass burning was the main source of NH_3, with other sources (wastes, soil, biogenic) predominant during the wet season. The estimated emission fluxes of NO_2-N, NH_3-N, NO_3^--N and NH_4^+-N from sugar cane burning in a planted area of about $2,2 \times 10^6$ ha were 11,0, 1,1, 0,2 and 1,2 Gg.N.yr^{-1}, respectively.

The sources, atmospheric transport and reactions of the main inorganic reactive nitrogen (N_r) species (NO_2, NH_3, HNO_3 and aerosol nitrate and ammonium) were investigated in a study conducted over a period of one year at six sites distributed across an area of about 130,000 km^2 in São Paulo State [11]. Oxidized forms of nitrogen were estimated to account for about 90% of dry deposited N_r, due to the emissions of nitrogen oxides from biomass burning and road transport. NO_2-N was important closer to urban areas; however, HNO_3-N was the largest individual component of dry deposited N_r. A simple mathematical model was developed to enable determination of total N_r dry deposition from knowledge of NO_2 concentrations. The model, whose error ranged from <1% to 29%, provided a new tool for the mapping of reactive nitrogen deposition.

The sugar cane burning emissions radically alter the chemistry of precipitation water. Coelho et al. [122] found that concentrations of soluble ions (K^+, Na^+, NH_4^+, Ca^{2+}, Mg^{2+}, Cl^-, NO_3^-, SO_4^{2-}, F^-, PO_4^{3-}, CH_3COO^-, $HCOO^-$, $C_2O_4^{2-}$ and HCO_3^-) increased by between two and six-fold during the harvest period. Principal component analysis revealed three main sources of the material in rainwater: biomass burning and soil dust re-suspension (52% of the total variance), secondary aerosols (26%) and vehicular emissions (10%). The biomass burning component diminished in the summer (non-burning period), when there was a relative increase in the importance of road transport/industrial emissions. The volume-weighted mean concentrations of ammonium (23,4 µmol.L^{-1}) and nitrate (17,5 µmol.L^{-1}) in rainwater samples collected during the harvest period were similar to those found in rainwater from São Paulo city, which emphasized the importance of including rural agro-industrial emissions in regional-scale atmospheric chemistry and transport models. There was evidence of a biomass-burning source throughout the year, which suggests that vegetation fires may continue to emit aerosols and their precursor gases, even after sugar cane burning is phased out.

4.2. Metropolitan São Paulo (RMSP)

In terms of trace species, the composition of the lower troposphere in the conurbation of the RMSP differs considerably from that of the interior of the State and the coastal zone. The critical air quality issue here is the scale of the emissions from road vehicles. In 2001, the vehicle fleet consisted of 17,2% hydrated ethanol-fuelled, 76,3% gasohol-fuelled and 6,5% diesel-fuelled vehicles, with ethanol contributing 34% of the total fuel consumption [31]. The figures for 2011 were 46,7% gasohol (cars and light commercial), 3,9% hydrated ethanol, 31,9% flex-fuel, 5,4% diesel and 12% motorcycles [9].

It is important to consider the relative amounts of the different fuels used, since emissions vary according to fuel, which has consequences for aerosol composition. For example, there is a larger fraction of oxygenated compounds in the secondary aerosols produced from reactions involving the aldehydes and alcohols emitted during ethanol combustion, which can affect the hygroscopicity of the particles, as well as their toxicological properties [48, 49].

The proportions of gasohol (gasoline with 22% anhydrous ethanol) and hydrated ethanol used have varied considerably in recent decades. Ethanol was first adopted as a road vehicle fuel in Brazil in 1979, due to the Brazilian National Alcohol Program (PROALCOOL), which was introduced as a response to the 1970s oil crisis. This not only reduced Brazil's dependency on oil imports, but also helped to eliminate the use of lead-containing anti-knock additives [49]. Sales of hydrated ethanol-fuelled vehicles peaked in the 1980s [50]. More recently, since around 2005, the new car market has been dominated by flex-fuel vehicles equipped with engine systems able to adjust to the gasoline/ethanol mixture present in the fuel tank [9].

In 2011, the sources of PM_{10} in metropolitan São Paulo were: heavy goods vehicles (38,6%), re-suspended dusts (25%), secondary aerosols (25%), industrial processes (10%) and light duty vehicles (1,4%). Annual mean concentrations of PM_{10} measured at the 18 automatic monitoring stations in São Paulo ranged between 31 and 50 $\mu g.m^{-3}$ [9]. A detailed analysis of these measurements, as well as of $PM_{2.5}$, TSP and black smoke measurements made at a smaller number of locations, are provided in the CETESB report [9] and in earlier annual reports published by CETESB.

The pollutant source profile remains fairly constant throughout the year. Use of absolute principal factor analysis showed that the contributions of different sources to $PM_{2.5}$ mass during winter and summer were: vehicle emissions (28 and 24% for the two seasons, respectively), re-suspended soil dusts (25 and 39%), oil combustion (18 and 21%), sulphates (23 and 17%) and industrial emissions (5 and 6%). Soil dusts accounted for 75-78% of the mass of coarse particles [51]. Andrade *et al.* [14] reported the results of elemental analyses, using particle-induced X-ray emission (PIXE) analysis of fine and coarse aerosols collected in 1989. Principal component analysis revealed the following sources of fine particles: oil and diesel combustion (explaining 41% of the mass), re-suspended soil dusts (18%), industrial emissions (13%), and a source associated with emissions of Cu and Mg (18%). Sources of coarse particles were: re-suspended soil dusts (59%), industrial emissions (19%), oil burning (8%) and marine aerosols (14%). Alonso *et al.* [52] used chemical mass balance (CMB) receptor modeling to show that the composition of fine particles was consistent with the presence of primary material from vehicles and secondary organic carbon and sulphate. Road dust re-suspension and vehicle emissions were the main sources of coarse particles and TSP. The same trends in source profiles were observed at geographically distinct locations in São Paulo. Sanchez-Ccoyllo and Andrade [53] used receptor modeling to identify five main sources of aerosols: vehicles, waste incineration, vegetation, suspended soil dust and fuel oil burning.

Organic and elemental carbon, emitted mainly from diesel vehicles, together with ammonium sulphate, make up most of the mass of fine particles [54, 55]. In [56] it is reported that 80% of the mass of fine ($PM_{2.5}$) particles consisted of organic material, with SO_4^{2-}, NO_3^{-} and NH_4^{+} present in the fine fraction, and NO_3^{-}, SO_4^{2-}, Ca^{2+}, and Cl^{-} predominant in coarse particles ($PM_{2.5-10}$). Albuquerque *et al.* [57] found that fine particles were rich in BC, S and Pb, while elements associated with crustal aerosols and/or industrial emissions (Al, Si, Ca, Ti, and Fe), together with ammonium sulphate and BC, composed the coarse mode particles. Other species, including K, Al, Fe and soil minerals, are included as a smaller component of fine particle mass [46]. Both vehicular and industrial emissions are sources of trace metals (Zn, Pb, Cr, Mn, Cd, etc.) [58, 59], and there appear to be continuing emissions of Pb from the road vehicle fleet, despite apparently low levels of Pb in fuels [60].

Aerosol composition similar to that of São Paulo is found in other major conurbations. In Campinas, the second largest city in the State, 100 km inland from São Paulo, fine particles were found to consist of 48% elemental carbon and 22% organic carbon, together with soluble ions and trace elements [61].

The PM concentrations are influenced not only by the magnitudes of emission sources, but also by ventilation and relative humidity. Miranda and Andrade [54] reported that higher PM_{10} concentrations (105 $\mu g.m^{-3}$) measured during the winter of 1999, compared to winter 2000 (60 $\mu g.m^{-3}$), were due to both better ventilation of the city during the latter period, as well as an increase in particle sizes at higher humidity. Similar findings were reported in [53], with lower pollutant levels associated with increased ventilation, precipitation, and relative humidity.

Primary emissions from vehicles result in large diurnal cycles in the concentrations of PM_{10}, BC, CO, NO_x and SO_2 [51], however the diurnal trends in particle mass concentrations differ between highly polluted and less polluted periods, with concentrations higher during the daytime for the former, and during the nighttime for the latter [57]. A possible influence of humidity on both the mass and size distribution of the Sao Paulo aerosol was suggested by the observation that while the size distribution of ammonium sulphate was unimodal during the daytime (with a maximum at 0,38 μm), at night, when humidity is higher, the size distribution was bimodal (with maxima at 0,38 and 0,59 μm) [55]. Furthermore, particle growth, observed using a Scanning Mobility Particle Sizer (SMPS), has been found to increase under polluted conditions [57].

Although local sources are by far the most important contributors to particulate air pollution in São Paulo city, back-trajectory analysis has shown that the atmosphere of the city can also be affected by the advection of air masses from distant regions where agricultural biomass burning is practiced, especially northeast Brazil [62]. This could explain the finding that the relative contribution of ammonium sulphate is higher under less polluted conditions [57].

An important consequence of the prevalence of fine mode particles in the atmosphere of the city is that the indoor environment provides little or no protection against exposure to these pollutants, since they easily infiltrate buildings. This was observed [63] using simultaneous indoor and outdoor measurements of a range of ionic species associated with both primary emissions (potassium, magnesium, sodium and calcium) and secondary aerosol formation (chloride, acetate, nitrate, formate, pyruvate, nitrite, sulphate, oxalate and ammonium). The measurements were made in offices, restaurants and a hotel. In the fine mode, only oxalate and ammonium showed significantly lower concentrations indoors. In the coarse mode, lower concentrations were normally found indoors (with the exception of acetate, chloride and potassium), reflecting the less efficient infiltration of larger aerosols.

Polycyclic aromatic hydrocarbons are an important component of the urban aerosols. Chrysene, benzo(e)pyrene and benzo(b)fluoranthene were found to be the predominant PAHs in PM_{10}, originating from industry, vehicles and long-range transport [64]. Levels of PM_{10} similar to those in São Paulo were measured in a city (Araraquara) situated in the rural biomass burning zone, although here PAH concentrations were lower. In both cases, dry deposition appeared to be the main mechanism of removal of PAH-containing aerosols from the atmosphere [65].

Bourotte et al. [66] measured the concentrations of 13 PAHs in fine ($PM_{2.5}$) and coarse ($PM_{2.5-10}$) aerosols. In both fractions, the predominant compounds were indeno(1,2,3-cd)pyrene, benzo(ghi)perylene and benzo(b)fluoranthene and PAH ratios suggested that automobile exhaust was the main source of the compounds. Factor analysis revealed four source components for the $PM_{2.5}$ fraction: diesel emissions, stationary combustion, vehicle emissions, and combustion of natural gas and biomass. For the coarse fraction, two components were identified, corresponding to vehicles and a mixture of gas, oil, and waste combustion.

4.3. Coastal regions

Although measurements of atmospheric aerosol are scarce in most of the coastal regions, an exception is the industrialized town of Cubatão, located near sea level at the base of the Serra do Mar scarp, where there is a large industrial complex comprising over 20 heavy industries (petrochemical, chemical, iron and steel, fertilizer, cement, coking and others). The monitoring stations in this area register regular episodes of particulate pollution, with the emissions from the industrial installations being entrained into a sea breeze circulation, when PM_{10} concentrations can increase by as much as an order of magnitude [67]. Pollutants absorbed into cloud water and precipitation are subsequently deposited to the vegetation of the Serra do Mar Atlantic rainforest, causing extensive ecological damage [68].

Due to extreme levels of pollution, air quality in the Cubatão region has been monitored by CETESB since the 1980s, and there are currently three sites where PM_{10} is continuously measured (Figure 5), and one where TSP is measured [9]. The case of Cubatão is unique, since in contrast to the RMSP, by far the largest source of particulates is industrial emissions, rather than road transport. Guideline levels of TSP and PM_{10} have been frequently exceeded in the industrial zone (Vila Parisi) of Cubatão, and there has been no improvement in PM_{10} levels in recent years. During 2011, the annual mean PM_{10} concentrations in the three zones of Cubatão were 99 μg.m^{-3} (Vila Parisi), 61 μg.m^{-3} (Vila Mogi) and 38 μg.m^{-3} (Centro) [9]. At the industrial Vila Parisi site, the annual geometric mean TSP concentration was 236 μg.m^{-3}, greatly exceeding the primary and secondary air quality standards for this pollutant species (80 and 60 μg.m^{-3}, respectively).

Although industrial emissions are responsible for the largest proportion of the aerosol loading of the atmosphere near the Cubatão industrial complex, the organic fraction has an important road transport-related component, because concentrations of polycyclic aromatic hydrocarbons (PAHs) are governed by emissions from heavy duty diesel vehicles [69]. In the same work, it was reported that a shift to greater use of biodiesel might decrease emissions of the PAHs.

In regions distant from the industrial installations, the aerosol composition reflects mainly natural sources (biogenic, terrigenous and marine). Bourotte *et al.* [70] found that aerosol (PM_{10}) composition in a State Park in the Cunha region was characterized by an abundance of K^+, Ca^{2+}, Na^+, Cl^- and Pb, while Vasconcellos *et al.* [71] reported the presence of aliphatic hydrocarbons emitted from biogenic sources in the coastal region.

5. Lidar observations

5.1. MSP-Lidar

In 2001, an elastic backscattering lidar system (MSP-Lidar) was installed in a suburban area of São Paulo city, on the Campus of the University of São Paulo (23°33′ S, 46°44′ W; Figure 6) and is being operated by the *Centro de Lasers e Aplicações* (CLA) of the *Instituto de Pesquisas Energéticas e Nucleares* (IPEN). The lidar is collocated with an AERONET sunphotometer, which provides the vertical profile of the aerosol backscatter coefficient at 532 nm up to an altitude of 4–6 km above sea level [72]. The MSP-Lidar comprises a Nd:YAG laser with a wavelength of 532 nm, and is operated with a repetition rate (PRF) of 20 Hz and an energy pulse of up to 120 mJ. The backscattering signal is captured by a Newtonian telescope with 1,5 m focal length. Attached to the telescope is a photomultiplier optimized for the visible spectrum with a 1 nm FWHM interference filter. Observations are being made whenever atmospheric conditions (absence of low or middle clouds; no rain) permit the operation of the lidar, resulting in a vast amount of data having been accumulated, which have so far been exploited in 5 MSc and 4 PhD theses, the most relevant being [73-76].

In January 2004, the IPEN MSP-Lidar system was installed for 6 weeks at IPMet in Bauru (Figure 6), located in the central part of São Paulo State, to provide the first measurements of aerosol layers in the interior of the State [77]. At the beginning of the campaign, the lidar was operated in its original configuration and the data were digitized using a digital oscilloscope with 1 GHz bandwidth and 11-bit resolution; at the end of January 2004, this device was replaced by a transient recorder, capable of simultaneous analog and photon counting measurements at higher resolution (12-bit). The system was operated on 31 different days, during periods of about 4 hours in the morning, 4 hours in the afternoon and 6-8 hours during the night, depending on the occurrence of cloud and/or precipitation. The daytime measurements had a 15-30 m spatial resolution and maximum altitude of 10 km, yielding information on the diurnal variation of the Planetary Boundary Layer (PBL), while the measurements at night had a 30-60 m resolution, reaching up to 30-35 km maximum altitude. The diurnal variation of the PBL during the austral summer could be documented, as well as some background concentrations of aerosols, because very little biomass burning takes place during the rainy period.

Figure 10 shows a typical example of the diurnal variation of the height of the PBL on a cloudless day in Bauru. It should be noted that, due to the latitude of -22,3°, the lidar cannot be operated during the midday period in summer, but as a result of turbulent mixing, the PBL could easily reach a maximum height of ≥3,5 km above ground level (AGL) during the early afternoon. The top of the PBL starts decreasing well before sunset, until it stabilizes at around 1,5 km AGL during the night. Times are indicated in Local Time (LT = UT-3h).

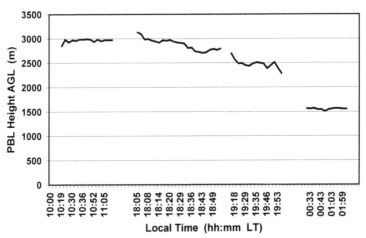

Figure 10. Height of the PBL over Bauru on 01/02 March 2004 for four different periods (10:19-11:08; 18:05-18:55; 19:13-19:54; 00:27-02:02 LT), with a vertical resolution of 30 m (after [78]).

The increased vertical range of the lidar during nocturnal operation permitted the detection of thin clouds and layers of aerosols, as shown in Figure 11. A cloud layer is clearly visible at around 4,5 km, while aerosols were detected at 3 and 5 km AGL, respectively. The top of the PBL is at about 1850 m AGL, with the faint layering being indicated in shades of green and light-blue colours.

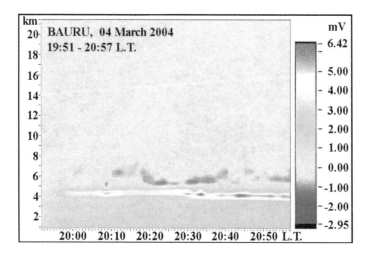

Figure 11. Nocturnal lidar observation above Bauru on 04 March 2004. Vertical range from 855 m to 21,5 km AGL, with a resolution of 30 m (after [78]).

In early 2008, the MSP-Lidar system was upgraded to a Raman lidar, and in its present 3-channel configuration it can measure elastic backscatter at 355 nm, together with nitrogen and water vapour Raman backscatters at 387 nm and 408 nm, respectively. Therefore, the PBL data now available include aerosol backscattering and extinction coefficients, as well as the Lidar Ratio (LR) and water vapour mixing ratio. Figures 12 and 13 present typical results of Raman lidar measurements recorded during night-time of 09/10 January 2008, during the austral summer season. This period of the year is characterized by a very well defined boundary layer throughout the day and relatively high humidity. The major part of aerosols and water vapour is contained within the boundary layer, while the scattering above the PBL is mainly due to molecules. Figure 12 shows the aerosol extinction and backscattering coefficient profiles at 355 nm, where one can see a residual aerosol layer between 900 m and 2000 m AGL, indicating a very pronounced presence of aerosols, overlaid by another discrete layer above it between 2500 m and 3500 m AGL. The height profile of the Lidar Ratio is shown in Figure 13a. The Lidar Ratio is about 80 sr and stable throughout the PBL up to about 3000 m AGL. The vertical profile of the lidar-derived water vapour mixing ratio can be seen in Figure 13b. The calibration of the lidar was performed using radiosonde data from the nearby São Paulo Campo de Marte airport. Although the sonde had a relatively low height resolution, integrating the water vapour content with height made such calibration possible.

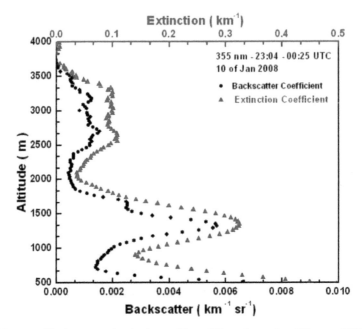

Figure 12. Aerosol backscatter and extinction profiles at 355 nm observed on 10 January 2008 at 00:25 UT (21:25 LT). The PBL top height is considered to be at 2000 m.

The MSP-Lidar system has contributed to several studies concerning the properties of aerosols and their influence on the air quality index of the city of São Paulo. Lidar measurements conducted daily provided observations of the PBL variation, which could be compared to corresponding air quality index values from local air quality monitoring and management agencies, as well as identifying potential air dispersion conditions [79]. It has also been deployed to monitor the long-range transport of aerosol plumes from different regions of Brazil to the RMSP and to evaluate the contribution of aerosol pollutants from remote sources. Landulfo and Lopes [80] have analyzed an event during the period 02 - 09 August 2007 when the AOD (Aerosol Optical Depth) and AE (Ångström Exponent) values retrieved from the AERONET sunphotometer indicated that high aerosol loads at five different locations in the Brazilian territory corresponded to biomass-burning particles. This was validated by the mean values of the Total Attenuated Backscatter Coefficient at 532 nm, the mean depolarization ratio and also the Lidar Ratio (about 70 sr) for all sites over-flown by the CALIOP sensor onboard the CALIPSO satellite.

In another case study during the dry winter season of 2008, fire plumes attributed to sugar cane fires were frequently observed by IPMet's radars in the absence of rain echoes and documented in terms of radar reflectivity, time and location [12]. On several occasions, IPEN's Elastic Backscatter Lidar in São Paulo observed layers of aerosols of varying strength and heights above the city. The most significant days were selected for calculating

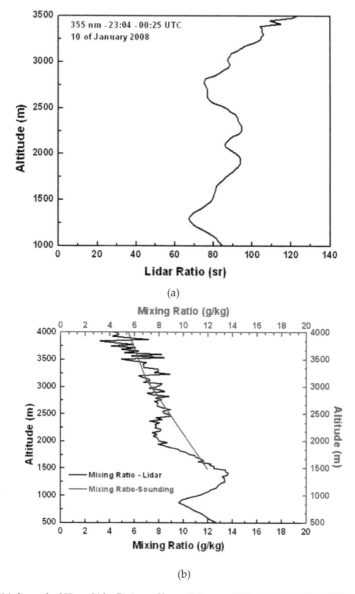

(a)

(b)

Figure 13. (a) shows the 355 nm Lidar Ratio profile on 10 January 2008 at 00:25 UT (21:25 LT).   **(b)** shows the water vapour mixing ratio extracted on the same day from the 408 nm channel (00:25 UT) and from a radiosonde ascent (00:00 UT).

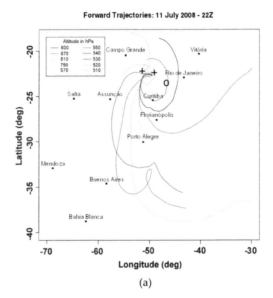

(a)

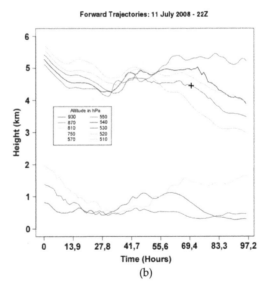

(b)

Figure 14. Forward trajectories initiated at different heights where a large fire was observed by IPMet's radars on 11 July 2008, 22:00 UT (19:00 LT). **(a)** The + indicates the position of the PPR and BRU radars; o indicates the position of the lidar in São Paulo (IPEN). **(b)** Forward trajectories plotted against height and time. The + indicates the position of IPEN, marking height and time of arrival matching exactly with the lidar observation (Figure 15).

backward, as well as forward trajectories, deploying the Flextra 3.3 Trajectory Model [81], which was initiated with ECMWF historical data with a 0,25° x 0,25° grid spacing [12]. The results showed an excellent match between the radar-detected sources of the plumes and lidar observations in São Paulo. Figure 14 presents a typical case study, when emissions from biomass fires were identified by the radars on 11 July 2008 in the central parts of the State, and were subsequently monitored by IPEN's lidar over Metropolitan São Paulo on 14 July 2008, deploying forward and backward trajectories. The forward trajectories, initiated at different heights ranging from 930 hPa (close to ground level) up to 450 hPa (ca 6,7 km amsl) at 30 hPa intervals (only the most significant 10 heights are shown in Figure 14), indicated a transport duration of approximately 70 hours under the prevailing meteorological conditions (Figure 14b). The arrival of the plume over the RMSP on 14 July 2008, as observed by the lidar at IPEN, is shown in Figure 15.

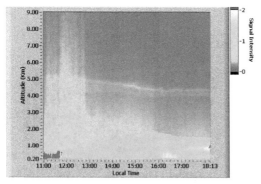

Figure 15. Lidar observations at IPEN in São Paulo, showing the range-corrected signal in arbitrary units, on 14 July 2008 between 11:00 and 18:13 LT. The plume identified in Figure 14 can be seen between 4-5 km AGL.

The MSP-Lidar system in São Paulo has also been contributing to CALIPSO satellite validation procedures [75, 82]. During 2007, correlative measurements were carried out with special attention to the dry season (May-October), when most of the days have poor dispersion conditions and long distance transport is more frequent. From a total of 28 days of measurements, on only 10 days were no clouds present below 4 km. Figure 16a presents a typical example, showing the range-corrected signal retrieved by the lidar system at São Paulo on 10 October 2007 between 03:34 and 05:35 UT, which contains the CALIPSO overpass window, beginning at 04:30 UT (Figure 16b). On this day, the closest distance of the satellite ground-track from the lidar site was about 48 km. The presence of aerosol layers above the PBL at 4-5 km, 6 km and 9 km is noticeable. The same features are also observed in the CALIOP 532 nm Total Attenuated Backscattering plot, as shown in Figure 16b. Both systems detected a cirrus structure between 12 and 13 km AGL, but the strong cirrus cloud signal observed in the CALIOP "plot-curtain" is much weaker in the lidar image. The red box in Figure 16b represents the CALIPSO ground-track region over Metropolitan São Paulo with coordinates of -22,5625° latitude and -46,0247° longitude at about 04:35 UT.

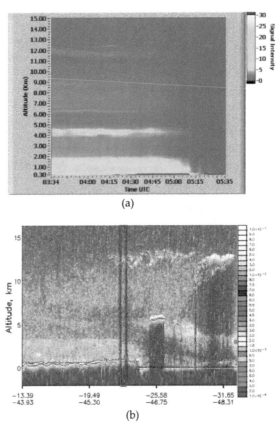

(a)

(b)

Figure 16. **(a)** Range-corrected lidar signal (plot-curtain) measured by the MSP-Lidar on 10 October 2007, 03:34 - 05:35 UT. **(b)** Total Attenuated Backscattering signal measured by the CALIOP at 532 nm during the period 04:30 - 04:41 UT on the same day, when it was closest to the MSP-Lidar site (red box).

Figure 17 compares the attenuated backscatter coefficient profile retrieved by CALIOP on board the CALIPSO satellite and the corrected one obtained from the ground-based MSP-Lidar system in São Paulo. The satellite profile has a 5 km horizontal resolution. The attenuated backscatter profile from the MSP-Lidar site was derived under cloud-free conditions from the range-corrected and background noise-subtracted lidar return signal. Both profiles are in good agreement, presenting similar layer patterns in the profiles observed at 5-6 km and about 7 km AGL. Since it can be assumed with reasonable confidence that, at higher altitudes, the horizontal atmospheric structure is more homogeneous, the good agreement between the two systems demonstrates the possibility that they were probing the same air masses for this specific measurement. At lower altitudes, observation of some differences between the two profiles is more likely due to local effects. In this case, the localized effects are more pronounced, and the fact that the systems are not covering the exact same region becomes evident.

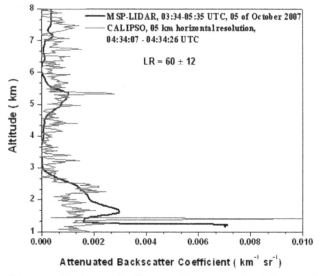

Figure 17. Total Attenuated Backscatter Coefficient profiles at 532 nm for the horizontal coverage of CALIPSO level 1 data compared to the Attenuated Backscatter Coefficient retrieved by the ground-based MSP-Lidar system in Metropolitan São Paulo on 10 October 2007.

5.2. Mobile Raman lidar

The mobile bi-axial Raman lidar system uses a commercial pulsed Nd:YAG laser, operating at a wavelength of 532 nm in the elastic channel and 607 nm in the Nitrogen Raman channel, with a pulse energy of 130 mJ at 20 Hz PRF. The pulse width is 25 ns, yielding a spatial resolution of 7,5 m. A detailed description of the system is found in [83]. The system allows the determination of the optical properties of the atmosphere, including aerosol backscatter and extinction coefficients, as well as an indication of the type of aerosol present, based on the Lidar Ratio. This lidar has so far been deployed during specific campaigns at three different sites within the central region of São Paulo State, *viz.*, Rio Claro [84], Bauru and Ourinhos [85-87], as well as in Cubatão, an industrial hub at the coast, near Santos [88], as shown in Figure 6.

A one-month pilot study was undertaken during August 2010 in Ourinhos (Figure 6), which is situated in one of the State's major sugar cane producing regions, where biomass burning is a regular occurrence. The objective was to characterize the effects of these emissions on the atmosphere, considering the local circulation and the consequences for the region [85]. In the absence of rain, the plumes were tracked by IPMet's two S-band Doppler radars within their quantitative ranges of 240 km (BRU = Bauru, PPR = Presidente Prudente; Figure 6), using the TITAN (*Thunderstorm Identification, Tracking, Analysis, and Nowcasting*) Radar Software [89]. A large range of meteorological, physical and chemical instrumentation, including the mobile Raman lidar, was used to observe elevated layers and the type of aerosols. A medium-sized sodar, as well as 6 automatic weather stations, were also deployed in the region. Various gases and aerosol size fractions were sampled, providing an atmospheric chemistry database and thus documenting the impact of the harvesting practice on the region. The aerosol load of the atmosphere was quantified by hourly mean AOD values and hourly mean backscatter profiles. Several case studies have already been analyzed, but the one of 25-26 August 2010 will be shown in this Chapter to illustrate how the various remote sensing instruments are being deployed to generate a complete picture of events.

During the second half of August 2010, the weather was dominated by a high pressure system, resulting in a rise in temperatures, with low humidity favoring the accumulation of pollutants in the atmosphere of the region [25]. IPMet's radars have a 2° beam width and a quantitative range of 240 km, generating a volume-scan every 7,5 minutes, with a resolution of 250 m radially and 1° in azimuth. Reflectivities and radial velocities are recorded at 16 elevations. However, in order to detect and track the biomass burning plumes, a special scanning cycle was configured to provide a better vertical resolution up to the anticipated detectable top of the plumes: 10,0°, 8,0°, 6,5°, 5,0°, 4,0°, 3,2°, 2,4°, 1,6°, 0,8° and 0,3°, with each "sweep" (Plan Position Indicator - PPI) having 360 rays with 957 range bins each. Two different software systems were deployed, viz., IRIS (*Interactive Radar Information System*) Analysis was used first to generate CAPPIs (Constant Altitude PPIs) at 1,5 and 2,0 km amsl, in order to identify all smoke plumes within the 240 km range of the radars. Once a plume was identified as likely to pass over the monitoring site, it was tracked using TITAN Software to determine its intensity (based on radar reflectivity in dBZ), horizontal and vertical dimensions, and the velocity of approach. The thresholds used for tracking were 10 dBZ with a minimum volume of 2 km^3. It should be noted that TITAN uses Universal Time (Local Time LT = UT-3h).

A typical case study of a sugar cane fire in the Ourinhos region is now presented, demonstrating the integration of all types of data into one coherent event. The first echo of a smoke plume was detected by the Bauru radar on 26 August 2010 at 00:08 LT, about 35 km north-northeast of Ourinhos and about 85 km southwest of the radar (Figure 18), rapidly gaining in area and intensity (≤40 dBZ near its origin). By 00:22 LT, the TITAN Software could already identify its centroid of ≥10 dBZ reflectivity and tracked it until 02:45 LT, when the plume had already spread over Ourinhos, where the Raman lidar and sodar were located. As the plume moved southwards with the northerly winds, the aerosols spread out (dispersed) and the reflectivity dropped gradually, but it could still be detected by the radar until 03:46 LT, >20 km south of Ourinhos, using a reflectivity threshold of –6 dBZ [85].

Furthermore, it can be deduced from Figure 18a that while the plume was at a low height during the initial phase of transport, it moved very slowly (3-4 km.h^{-1}), since the wind speed in the first few hundred meters was very low (≤5 m.s^{-1}), as observed by the sodar. There was also a shift of the wind direction from easterly to northerly winds above 300 m AGL. These northerly winds were above the nighttime surface inversion, confirmed by the "Skew T x Log P" profiles of the Meso-Eta model in the 900-800 hPa layer (650–1650 m AGL) as shown in [85]. The vertical velocity (w), measured by the sodar, indicated that downward mixing of the pollutants (aerosols), trapped above the inversion, only commenced at around 09:00-09:30 LT, since from 00:00-09:00 LT the atmosphere was extremely stable below 300 m AGL (w = ±0 m.s^{-1}).

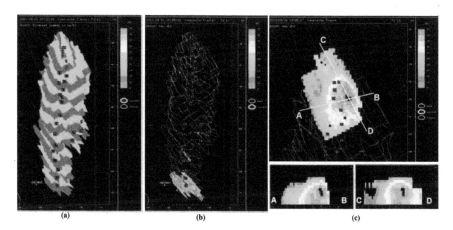

(a) First TITAN centroid of the *queimada* (actual fire, blue) at 03:22 UT (00:22 LT; annotation: propagation velocity in km.h^{-1});

(b) The *queimada* reached the Ourinhos region at 05:45 UT (02:45 LT, blue; annotation: maximum reflectivity in dBZ).

(c) Vertical cross-sections at 03:45 UT (00:45 LT), showing the horizontal and vertical extent along the base lines A-B and C-D.

Figure 18. Examples of the tracks generated by TITAN on 26 August 2010. The envelopes (10 dBZ reflectivity) show the position of the *queimada* (smoke plume) in intervals of 7,5 min (blue = actual time; green = future; yellow = past).

The lidar observed the arrival of the plume at 02:40 LT between 350 and 600 m AGL (Figure 19a). The top of the PBL extended to about 2,6 km AGL, above which a very dry and relatively warm and clean air mass was advected from the west, creating an elevated inversion which blocked further upward mixing. The lowest layer ≤250 m AGL appeared clean, being trapped within the surface inversion, inhibiting downward mixing, also confirmed by the sodar measurements, indicating a very stable layer. Lidar data from the Raman Channel (non-elastic signal at 607 nm) were integrated into hourly means until 09:00 LT to obtain the AOD. The results confirmed a high aerosol load of the atmosphere, with hourly mean values of AOD varying between 0,265 and 0,288 until 07:00 LT, after which they increased to 0,433 by 09:00 LT. Hourly means of the Lidar Ratio confirmed the arrival of the plume between 02:00 and 03:00 LT (example shown in Figure 19b), while an almost 20% increase of LR to 72 sr after 07:00 LT was probably due to downward mixing of the aerosols accumulated above the inversion, also confirmed by an increase of AOD values from the Raman signal [85]. LR values of around 70 sr suggest aerosols originating from biomass burning [90, 91].

Visual images from overpasses of the MODIS-AQUA satellite on 25 and 26 August 2010 (at 17:35 and 16:40 UT, respectively; 14:35 and 13:40 LT) showed intense smoke plumes to the west and south of the Ourinhos region, with AOD values of up to about 1,0. In the Ourinhos region, the AOD increased during the period 25-26 August, from about 0,2 to about 0,6 (Figure 20a), which is in agreement with the early afternoon lidar measurements (Figure 20b), which provided an AOD value of 0,380 during the period from 13:00 to 14:00 LT.

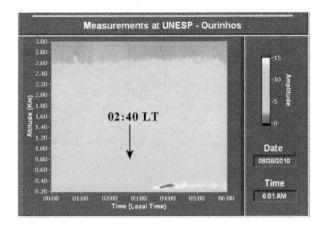

(a)

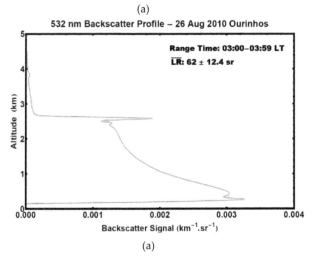

(a)

Figure 19. (a) Lidar signal (arbitrary units) visualized for 00:00-06:00 LT, up to 3 km AGL.
(b) Backscatter Profile at 532 nm for the hourly mean period 03:00-03:59 LT on 26 August 2010.

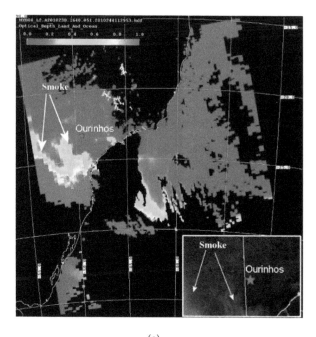

(a)

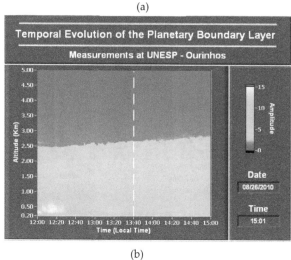

(b)

Figure 20. (a) AOD image from MODIS-AQUA on 26 August 2010, 16:40-16:45 UT (13:40-13:45 LT). The inset shows a simultaneous visual image of the Ourinhos region.
(b) Lidar measurements on 26 August 2010, 12:01-15:01 LT. The time of the MODIS-AQUA overpass is indicated by the dashed white line.

Aerosols collected during daytime and nighttime periods at the lidar site [85-87, 92], using low-volume filter samplers, were chemically characterized by means of ion chromatography. A higher concentration of K^+ during the period from 22:00 on 25 August to 16:00 on 26 August 2010 indicated the presence of biomass-burning material (Figure 21), since K^+ is a plant macronutrient released during the combustion process. Levoglucosan, a very specific chemical marker of biomass combustion, was well above average concentration during day sampling on 26 August and even higher during the following night, indicating a strong presence of biomass smoke on both days.

In the study region, ions such as magnesium (Mg^{2+}) and calcium (Ca^{2+}) are associated with the re-suspension of soil dust, which often accompanies biomass fires due to the intense updrafts created. On 26 August, concentrations of these species were higher during the daytime, due to the increased emissions from barren fields and unsealed roads associated with higher wind speeds (Figure 21).

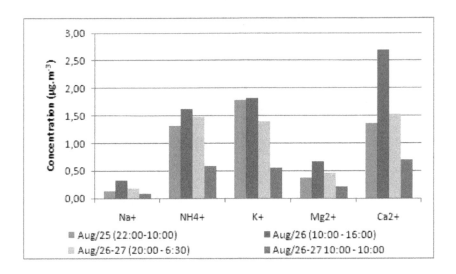

Figure 21. Soluble major cation concentrations for the period 25-27 August 2010 (sampling periods are indicated in local time; after [85]).

Further evidence of the impact on the Ourinhos region of emissions from sugar cane fires was obtained by comparing the concentrations of organic compounds in aerosol particles collected on 26 August with those collected one day earlier. Ambient levels of polycyclic aromatic hydrocarbons (PAH), as well as PAH derivatives, such as oxy-PAH, were significantly higher on 26 August 2010 than on the previous day, confirming that emissions from sugar cane fires affected the urban atmosphere of Ourinhos.

5.3. Scanning lidar in Cubatão

An elastic backscatter lidar system, with similar characteristics to the mobile lidar, was installed in 2011 at CEPEMA-USP (*Centro de Pesquisas em Meio Ambiente,* a Center for Environmental Research and Training, under the responsibility of the Universidade de São Paulo) in the Cubatão industrial area, with the ultimate goal of remotely monitoring industrial emissions. It also uses a commercial pulsed Nd:YAG laser, operating at three wavelengths (355, 532 and 1064 nm) with pulse energies of 100, 200 and 400 mJ, respectively, at 20 Hz PRF. A detailed description of the system and its location is found in [33]. The system allows the determination of the optical properties of the atmosphere, including aerosol backscatter and extinction coefficients, as well as an indication of the type of aerosol present, based on the Lidar Ratio. The lidar is co-located with a sodar / RASS system and an air quality monitoring station.

During May 2011, the system was deployed in a vertical pointing mode during an intensive field campaign. A 24-hour period was selected that demonstrated the complexity of the local situation, which is dominated by topographical effects and prevailing meteorological conditions [33]. Vertical profiles of the Backscatter Coefficient (BSC) and the Colour Ratio were calculated for 30-minute periods from 17:30 – 19:59 and 21:42 – 23:36 LT. The BSC was highest for all frequencies between 19:30 and 19:59 LT (Figure 22a), indicating a strong inflow of aerosols, while after 21:42 LT the BSC showed much lower values (Figure 22b), representing a relatively clean air mass. At the same time, the Colour Ratio between all frequencies increased significantly, indicating the presence of small particles, especially between 0,8 and 1,3 km AGL [33]. Ground-level observations of PM_{10} and $PM_{2.5}$ for the 24-hour period indicate that PM_{10} concentrations were almost twice as high as those of $PM_{2.5}$ until about 18:00 LT (Figure 23). During the same period, the sodar observed extremely low wind speeds from varying directions. However, this resulted in very stable PBL conditions, and a temperature inversion began to develop from 18:30 onwards, reaching its greatest depth and intensity at 21:30. Thereafter, it gradually dropped in height and began to erode, as the air flow from the interior intensified, until it totally dissipated by 01:00 LT [33], due to the katabatic warming of the descending northerly airflow, which then also reduced the aerosol concentrations at ground level (Figure 23). Figure 24a shows the development of the surface inversion at 20:00 LT, overlaid by warm air flowing from the interior, with simultaneous downward motion below 240 m AGL (Figure 24b), highlighting the complex interaction of meteorology and topography in this region. This situation clearly demonstrates the need for solid environmental impact studies *before* locating industrial developments, in order to avoid any negative health impacts in the local population due to the accumulation of pollutants.

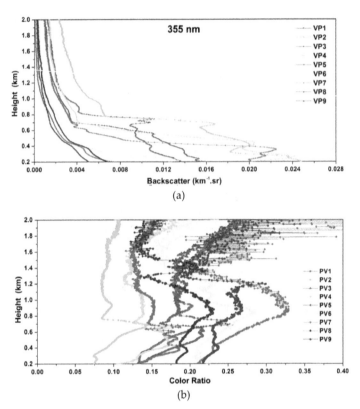

Figure 22. (a) Vertical profile of Backscatter Coefficient (BSC) at 355 nm; (b) Colour Ratio 532/355 nm. (After [33]).

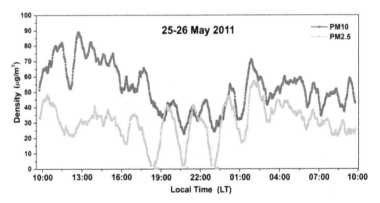

Figure 23. Concentrations of PM_{10} and $PM_{2.5}$ from 10:00 LT on 25 May to 10:00 LT on 26 May 2011. (After [33]).

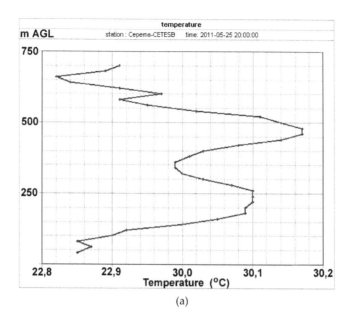

(a)

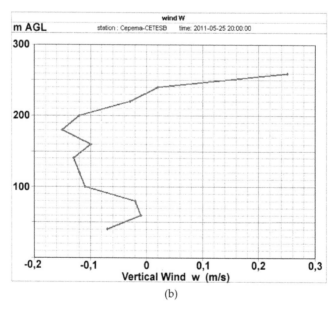

(b)

Figure 24. Sodar/RASS measurements on 25 May 2011, mean profiles for 19:30-20:00 LT. **(a)** Vertical profile of temperature; **(b)** Vertical profile of vertical wind velocity.

6. Impacts of aerosols

6.1. Impact of aerosols on human health

The impact of anthropogenic aerosols on human health has been acknowledged in both metropolitan and rural regions [93, and references therein]. In general terms, as pointed out in [93], in recent years the population of São Paulo State has suffered from either acute (short-term, high concentration) or chronic (long-term, lower concentration) exposure to particulate air pollutants, depending on location. In rural regions, there is acute exposure to high concentrations of biomass burning particulates present in plumes, as well as chronic exposure to these aerosols on a regional basis throughout the dry season. In metropolitan São Paulo, there is chronic exposure to particulates derived from road transport and industrial emissions, together with periodic acute exposure to extremely high levels of pollutants under conditions of thermal inversions and stationary air masses [94-96].

There have been many studies of the correlation between aerosol concentrations and human health impacts in the metropolitan regions, especially in São Paulo city [97-102]. Typical effects include asthma and pneumonia, as well as other cardiovascular and respiratory symptoms. Increased levels of PM_{10} were associated with increases of 6,7% and 2,2% in hospital admissions of children due to respiratory illness [96, 103]. Increments of 10 $\mu g.m^{-3}$ in PM_{10} concentrations resulted in increases in hospital admissions of between 0,9% and 6,7% in Sao Paulo [97, 103-105]. In the elderly, a 5,4% increase in the number of deaths was linked to a 10 $\mu g.m^{-3}$ increase in PM_{10} [105]. Industrial emissions in Cubatão have been found to seriously affect the lung function of children, with respiratory airflow rates correlated with PM_{10} concentrations obtained for the preceding month [106].

Bourotte *et al.* [56] investigated the relationships between peak expiratory flow (PEF) measurements and soluble ions in fine and coarse aerosols, and found a negative correlation between PEF and the coarse fraction ions Cl^-, Na^+, Mg^{2+} and NH_4^+, as well as between PEF and fine fraction Mg^{2+}. The findings suggested that increased levels of coarse particles could be of especial concern for asthmatic individuals.

In these heavily polluted regions, atmospheric particles contain components known to be carcinogenic and mutagenic, including ketones, aldehydes, quinolines, carboxylic acids, polycyclic aromatic compounds (PAHs), and nitro-PAHs. These substances have been associated with exhaust emissions from road vehicles in southeast Brazil [69, 107-109]. Benzo[a]pyrene equivalent values suggest that the cancer risk is greater for the São Paulo city aerosol than elsewhere in the State, although concentrations may not exceed World Health Organization guidelines [65].

Biomass burning emissions in rural regions also have a recognized influence on human health, as well as environmental impacts including modification of nutrient cycling [10, 11, 47], and effects on climate including alterations of the radiative properties of the lower atmosphere, cloud formation and precipitation [110, 111]. For these reasons, as well as due to the need to meet certification requirements of importing countries, there has been increased pressure for mechanization of harvesting, since the mechanized process does not necessarily require prior burning of the crop. Nonetheless, until recently burning has continued to be employed in mechanized areas (using simpler machinery) because it can improve economic efficiency by around 30-40% [50, 112].

A clear relationship between particulate air pollution and the occurrence of respiratory illness in sugar cane burning regions of the State has been reported [45, 113-116]. Particulate material from sugar cane burning was demonstrated to have the greatest detrimental effect on the respiratory systems of the most sensitive population groups. Cançado *et al.* [45] measured black carbon and trace elements in fine and coarse aerosol fractions, and related the concentrations to daily records of hospital admissions for respiratory illness of children (<13 years old) and the elderly (>64 years), in the town of Piracicaba. Increases of 10,2 μg.m^{-3} (PM$_{2.5}$) and 42,9 μg.m^{-3} (PM$_{10}$) were associated with increases in hospital admissions of 21,4% (children) and 31,03% (elderly people).

Carcinogenic and mutagenic compounds are emitted during biomass burning [109]. Concentrations of PAHs in a rural sugar cane burning region during the harvest period were in the range 0,5-8,6 ng.m^{-3} [38]. The mutagenic activity of PM$_{10}$ was much higher during the harvest season, when the PM$_{10}$ concentration was 67 μg.m^{-3}, and the mutagenic potency was 13,45 revertants m^{-3}. During the summer (non-burning period), the PM$_{10}$ concentration was 20,9 μg.m^{-3}, and the mutagenic potency was 1,30 revertants m^{-3} [117].

6.2. Impact of aerosols on rainfall

Aerosols derived from all of the sources described above are able to alter the radiative properties of the troposphere, and can modify the processes that lead to the development of cloud condensation nuclei, cloud droplets, and ultimately precipitation [118-120]. The magnitudes of these effects depend on the size distribution, number concentration and chemical composition of the particles, and can therefore vary widely within the same region.

Dufek and Ambrizzi [121] used daily precipitation data collected at 59 locations in São Paulo State to investigate rainfall trends for the period 1950-1999. Although some of the findings were contradictory, an overall trend towards a wetter climate was identified, with rainfall concentrated into a smaller number of more intense events. It was suggested that these changes could be related to the presence of biomass burning aerosols, as well as changes in land use. Evidence that the aerosols probably act as cloud condensation nuclei was provided in [122], where a relationship was identified between water-soluble organic carbon (WSOC) in the particles and dissolved organic carbon (DOC) in rainwater.

An important point is that sugar cane production in São Paulo State has increased over this period. Between the 1990/91 and 2000/01 seasons, the harvest increased from 132 Mt to 194 Mt [123]. It can therefore be supposed that there was also a large increase in emissions of aerosols from the burning of the crop, since manual harvesting of the cane (which requires prior burning) was the norm over the period. Mechanization of the crop (which does not involve burning) has only been introduced recently (from around 2005). The main conclusion to be drawn from this is that the trend towards a smaller number of more intense precipitation events, as reported in [121], could now be reversed in the interior of São Paulo State, as sugar cane burning is progressively phased out.

The climatological characterization of storm properties, such as area, volume, maximum echo top and reflectivity during two summer seasons, *viz.*, 1998-1999 and 1999-2000, based on observations from the Bauru S-band Doppler radar, has for the first time shown the spatial distribution of these parameters in central São Paulo State. Gomes and Escobedo [124] showed that some preferential areas of precipitation, taking into account a precipitation envelope area defined by the 25 dBZ threshold, were located along the Tietê River valley. The mean maximum reflectivity field (>40 dBZ), representing the cores of convective precipitation systems, has highlighted some preferential regions for convection to develop over urban and industrialized areas, such as metropolitan Campinas (Figure 25a). A climatology of flash density (Figure 25b; [21]) also identified Campinas as one of three regions with a higher concentration of lightning discharges, attributed to the occurrence of heat islands due to anthropogenic activities. Thus, the spatial distribution of the reflectivity field exceeding 40 dBZ in the Campinas region reinforces results showing a strong correlation between the frequency of cloud-to-ground lightning strokes and precipitation intensity.

The influence of anthropogenic aerosols on precipitation patterns in the region is the subject of research currently in progress as part of a thematic climate change research programme sponsored by FAPESP (the São Paulo State Research Foundation). The quantitative evaluation of changes in the rainfall pattern, such as increases or decreases of area rainfall totals, number and volume of convective cells, duration of rain events, distribution of echo top heights, etc., is in progress for a 10-year period of integrated radar observations (Bauru and Presidente Prudente radars), using the TITAN Software.

6.3. Impact of aerosols on the frequency of lightning

Westcott [125] documented for the first time an impact of large cities on the cloud-to-ground (CG) lightning frequency in the Midwest of the United States. This was followed up by various researchers around the world, including in Brazil [20] and ultimately summarized in [22], using 10 years of observations from the Brazilian Lightning Detection Networks (1999-2008). This research confirmed the impact of anthropogenic activities on lightning, but it also highlighted the complexity of the correlation between urban heat islands, concentrations of PM_{10} and SO_2 in terms of weekly cycles and meteorological conditions, such as CAPE (Convective Available Potential Energy) and other microphysical parameters. One of the most important findings was that the CG frequency increases with increasing concentrations of PM_{10} up to a certain threshold of PM_{10} concentration (saturation), after which it decreases with further increases of PM_{10} concentrations. As the CG frequency increases due to urban impacts, the percentage of positive strokes is reduced.

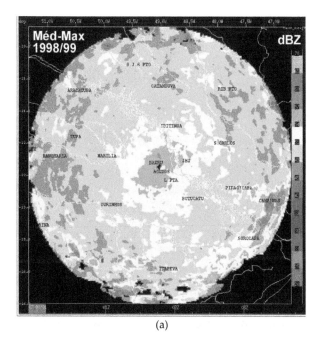

(a)

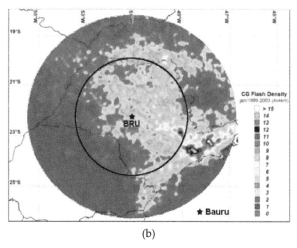

(b)

Figure 25. (a) Spatial distribution of the average maximum reflectivity (dBZ), during the period October 1998 to March 1999. TITAN storm threshold was defined as reflectivity >40 dBZ within the 240 km range of the Bauru Doppler radar (after [124]). **(b)** Flash density within the 450 km range of the Bauru radar (after [21]). The circle indicates the 240 km quantitative range as in the TITAN image above.

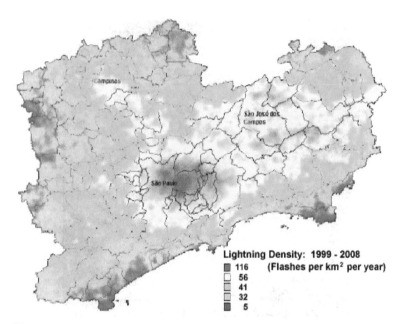

Figure 26. Lightning density (flashes per km² per year) during the period 1999 - 2008 for the eastern part of the State of São Paulo, which includes the major urban complexes, as well as the highly industrialized Paraiba Valley (after [22]).

7. Final considerations

This Chapter provides a review of all relevant historical data concerning the nature, concentrations and impacts of atmospheric aerosols in southeast Brazil. Highlights are the characterization of chemical, physical and optical properties of aerosols, as well as their geographic distribution within the State of São Paulo.

A significant reduction of mean annual PM_{10} concentrations could be noticed from 1998 onwards, confirming the success of the implementation of stringent air quality control measures, administered by CETESB. However, within the industrial suburb of Cubatão, confined in a valley, concentrations are still about twice the PQAr. After 2002, the annual mean PM_{10} concentrations in the RMSP and the interior of the state show relatively little year to year variation, but remain mostly below the Annual Standard (PQAr = 50 $\mu g.m^{-3}$).

At present, sugar cane burning, together with the re-suspension of soil dust that is inevitable during the harvesting process, is a major influence on aerosol concentrations, size distribution and dry deposition in rural regions of São Paulo State. However, in this region (and elsewhere in Brazil), the practice of pre-harvest burning is being eliminated. Recent legislation (State Law no. 11.241/02) envisages the complete cessation of the practice in mechanizable areas by 2021 and in non-mechanizable areas by 2031. Furthermore, an agreement between sugar cane producers and the State government has been reached, which involves elimination of burning in mechanizable areas by 2014, and in non-mechanizable areas by 2017 [9, 126]. This will have major environmental implications, including improvements in air quality and changes in the rates of deposition of nutrient species from the atmosphere to vegetation, soils and freshwater bodies [10]. Nonetheless, at present burning continues in 44% of the area planted with sugar cane [9]. Tsao *et al.* [127] suggest, using a life cycle analysis, that pollutant emissions in sugar cane regions are still increasing, due an expansion of the planted area, and that the burning step still contributes the largest fraction of the total emission.

Improvements in air quality in the metropolitan regions are likely to proceed at a slower pace than in the interior of the State, largely due to the dominant influence of emissions from the road transport sector. Nonetheless, emissions of aerosols and other pollutants are ultimately expected to be attenuated following progressive modernization of the vehicle fleet, and implementation of better controls on emissions from both vehicular and industrial sources.

Examples of case studies presented have demonstrated the capability of weather radars to detect, track and quantify emissions from biomass fires in the absence of rain echoes, deploying a special elevation scanning procedure to generate Volume-Scans every 7,5 min. Furthermore, satellites orbiting with lidar systems on board (e.g., MODIS-AQUA, CALIPSO, CloudSat) also have the capability to detect and quantify optical properties of aerosols.

With the gradual introduction of lidars in Brazil during recent years, it has also become possible to quantify in situ the vertical distribution and optical properties of suspended aerosols. However, in the State of São Paulo there are currently only three lidar systems available, *viz.*, one fixed lidar each in São Paulo city and in Cubatão, supplemented by the mobile lidar for periodic deployment in the interior of the State. Additional fixed lidar installations are therefore suggested for Campinas, Rio Claro, Bauru and São José do Rio Preto (situated in an important sugar cane production region in the north of the State) as a minimum configuration for a network, together with a second mobile system in São José dos Campos to cover the industrial activities in the Paraiba Valley.

Author details

Gerhard Held and Ana Maria Gomes
Instituto de Pesquisas Meteorológicas, Universidade Estadual Paulista, Bauru, S.P., Brazil

Andrew G. Allen and Arnaldo A. Cardoso
Instituto de Química, Universidade Estadual Paulista, Araraquara, S.P., Brazil

Fabio J.S. Lopes and Eduardo Landulfo
Centro de Lasers e Aplicações, Instituto de Pesquisas Energéticas e Nucleares,
Universidade de São Paulo, São Paulo, S.P., Brazil

8. References

[1] TRACE-A (1992). http://www-gte.larc.nasa.gov/trace/tra_hmpg.htm

[2] Lindesay J A, Andreae M O, Goldammer J G, Harris G, Annegarn H J, Garstang M, Scholes R J, van Wilgen B W (1996). International Geosphere-Biosphere Programme/International Global Atmospheric Chemistry SAFARI-92 field experiment: Background and overview, *J. Geophys. Res.*, 101, D19, 23,521-23,530.

[3] LBA (2012). http://lba.inpa.gov.br/lba/#

[4] Kocinas S, Artaxo P (1992). O monitoramento contínuo de elementos traços em aerossóis atmosféricos da bacia Amazônica. *Proceedings, VII CBMET*, São Paulo, 901-903. http://www.cbmet.com/cbm-files/19-02759b1503f7d21b92c9a5774714f2ba.pdf

[5] Allen A G, Cardoso A A, Da Rocha G O (2004). Influence of sugarcane burning on aerosol soluble ion composition in Southeastern Brazil. *Atmos. Environ.*, 38, 5025-5038.

[6] Da Rocha G O, Allen A G, Cardoso A A (2005). Influence of agricultural biomass burning on aerosol size distribution and dry deposition. *Environ. Sci. Technol.*, 39, 5293-5301.

[7] Lara L L, Artaxo P, Martinelli L A, Victoria R L, Ferraz E S B (2005). Characteristics of aerosols from sugar-cane burning emissions in Southeastern Brazil. *Atmos. Environ.*, 39, 4627-4636.

[8] IBGE (2011). Levantamento Sistemático da Produção Agrícola. Instituto Brasileiro de Geografia e Estatística (IBGE), Rio de Janeiro, December 2011: ISSN 0103-443X.

[9] CETESB (2012). Relatório 2011: Qualidade do Ar no Estado de São Paulo. CETESB, São Paulo, 124p.http://www.cetesb.sp.gov.br/ar/qualidade-do-ar/31-publicacoes-e-relatorios

[10] Allen A G, Cardoso A A, Wiatr A, Machado C M D, Paterlini W C, Baker J (2010). Influence of intensive agriculture on dry deposition of aerosol nutrients. *J. Braz. Chem. Soc.*, 21, 87-97.

[11] Allen A G, Machado C M D, Cardoso A A (2011). Measurements and modeling of reactive nitrogen deposition in southeast Brazil. *Environ. Pollut.*, 159, 1190-1197.

[12] Held G, Landulfo E, Lopes F J S, Arteta J, Marecal V, Bassan J M (2011). Emissions from sugar cane fires in the central & western State of São Paulo and aerosol layers over metropolitan São Paulo observed by IPEN's lidar: Is there a connection? *Opt. Pura Apl.*, 44, 83-91.

[13] Orsini C M Q, Artaxo P (1983). Algumas caracteristicas da materia particulada inalavel em Cubatao. In: *Seminário sobre uma Sintese do Conhecimento sobre a Baixada Santista*, CETESB, São Paulo, SP, 1983, 516p.

[14] Andrade M F, Orsini C, Maenhaut W (1994). Relation between aerosol sources and meteorological parameters for inhalable atmospheric particles in Sao Paulo City, Brazil. *Atmos. Environ.*, 28, 2307-2315.

[15] Wikipedia (2011). http://en.wikipedia.org/wiki/Largest_cities_in_the_world (Accessed in December 2011).

[16] IBGE (2010). http://www.censo2010.ibge.gov.br/resultados_do_censo2010.php (Accessed in December 2011).

[17] Saldiva P H N, Pope C A, Schwartz J, Dockery D W, Lichtenfels A J, Salge J M, Barone I, Bohm G M (1995). Air pollution and mortality in elderly people: A time-series study in São Paulo, Brazil. *Arch. Environ. Health*, 50, 159–163.

[18] RINDAT (2012). http://www.rindat.com.br/

[19] ELAT/INPE (2012). http://www.inpe.br/webelat/homepage/

[20] Naccarato K P, Pinto Jr O, Pinto I R C A (2003). Evidence of thermal and aerosol effects in the cloud-to-ground lightning density and polarity over large urban areas in Southeastern Brazil - Overview and Comparison to the Campaign Period. *Geophysical Res. Letters*, 30, 13, 1674-1677, doi: 10.1029/2003GLO17496.

[21] Naccarato K P, Pinto Jr O, Held G, 2004. Climatology of Lightning in Brazil - Overview and Comparison to the Campaign Period. *Proceedings, HIBISCUS / TroCCiBras / TROCINOX Workshop*, Bauru, SP, 16-19 November 2004, p10.
http://www.ipmet.unesp.br/ipmet_html/troccibras/publicacoes.html

[22] Gomes W R (2010). Estudo das caraterísticas da atividade dos raios na região metropolitana de São Paulo. PhD thesis, Geofísica Espacial, INPE, São José dos Campos, Brazil, 157p.

[23] INMET (2012). http://www.inmet.gov.br/html/clima/mapas/?mapa=prec

[24] Kousky V E (1988). Pentad outgoing longwave radiation climatology for the South American sector. *Rev. Bras. Meteor.*, 3, 217-231.

[25] Held G, Nery J T, Gomes A M, Lopes F J S, Ramires T, Lima B R O (2012). Study of biomass emissions in the central State of São Paulo: meteorological conditions during August 2010 cause an accumulation of pollutants in the Ourinhos region. *Proceedings, XI Congresso Argentino de Meteorologia*, Mendoza, Argentina, 28 May – 01 June 2012, 8p.

[26] Held G, Bassan J M, Frascarelli Jr R S (2011). Continuous Monitoring of the Lower Boundary Layer in the central State of São Paulo, Brazil, with a SODAR. *Geophysical Research Abstracts*, 13, EGU General Assembly 2011, Vienna, Austria, 03-08 April 2011. http://presentations.copernicus.org/EGU2011-13634_presentation.pdf

[27] Feliz G S (2012). Aspectos sobre a análise meteorológica do jato de baixo nível na cidade de Bauru. Monografia para BSc (Geografia), USC, Bauru, 2012, 70p.

[28] Karam H A (2002). Estudo do Jato de Baixo Nível de Iperó e das Implicações no Transporte de Poluentes no Estado de São Paulo. Tese de doutorado, IAG/USP, São Paulo, 2002, 213p.

[29] Held G, Danford I R, Hong Y, Tosen G R, Preece A R (1990). *The life cycle of the low-level wind maximum in the Eastern Transvaal Highveld: A cross-sectional study*. A Report to the CSIR Executive and the Management of Eskom Engineering Investigations, CSIR Report EMA-C 90146, Pretoria, September 1990, 50p.

[30] CETESB (2012).
http://sistemasinter.cetesb.sp.gov.br/Ar/mapa_qualidade/mapa_qualidade_interior.asp?id=350431

[31] CETESB (2002). Relatório de qualidade do ar do Estado de São Paulo em 2001. CETESB, São Paulo, 132p. http://www.cetesb.sp.gov.br/ar/qualidade-do-ar/31-publicacoes-e-relatorios

[32] WHO (2005). *WHO Air Quality Guidelines: Global update 2005*. Report on Working Group Meeting, Bonn/Germany, 18-20 October 2005, 30p.

[33] Steffens J, Da Costa R F, Landulfo E, Guardani R, Moreira P F Jr, Held G (2011). Remote sensing detection of atmospheric pollutants using lidar, sodar and correlation with air quality data in an industrial area. *Proceedings, SPIE Remote Sensing Conference*, Prague, Czech Republic, 19-22 September 2011, v.8182, 81820Z; doi: 10.1117/12.897915.

[34] Da Rocha G O, Allen A G, Cardoso A A (2004). Influence of sugar cane burning on aerosol soluble ion composition in southeastern Brazil. *Atmos. Environ.*, 38, 5025-5038.

[35] Campos M L A M, Urban R C, Da Silva L C, Souza M L, Allen A G (2012). Use of levoglucosan, potassium, and water-soluble organic carbon to characterize the origins of biomass burning aerosols. Submitted to *Atmos.Environ.*, March 2012.

[36] Da Rocha G O, Franco A, Allen A G, Cardoso A A (2003). Sources of atmospheric acidity in an agricultural-industrial region of São Paulo State, Brazil. *J. Geophys. Res.*, 108(D7), 1-11.

[37] Godoi R H M, Godoi A F L, Worobiec A, Andrade S J, de Hoog J, Santiago-Silva M R, Van Grieken R (2004). Characterisation of sugar cane combustion particles in the Araraquara region, *Southeast Brazil. Microchim. Acta*, 145, 53-56.

[38] Godoi A F L, Ravindra K, Godoi R H M, Andrade S J, Santiago-Silva M, Van Vaeck L, Van Grieken R (2004). Fast chromatographic determination of polycyclic aromatic hydrocarbons in aerosol samples from sugar cane burning. *J. Chromatog.*, A 1027, 49-53.

[39] Kirchhoff V W J H, Marinho E V A, Dias P L S, Pereira E B, Calheiros R, André R, Volpe C (1991). Enhancements of CO and O3 from burnings in sugar cane fields. *J. Atmos. Chem.*, 12, 87-102.

[40] Oppenheimer C, Tsanev V I, Allen A G, McGonigle A J S, Cardoso A A, Wiatr A, Paterlini W, Dias C M (2004). NO2 emissions from agricultural burning in São Paulo, Brazil. *Environ. Sci. Technol.*, 38, 4557-4561.

[41] CETESB (2002). Companhia de Tecnologia de Saneamento Ambiental. Avaliação dos compostos orgânicos provenientes da queima de palha de cana-de-açúcar na região de Araraquara e comparação com medições efetuadas em São Paulo e Cubatão (relatório final – 2002). CETESB, São Paulo, 97p. http://www.cetesb.sp.gov.br/Ar/publicacoes.asp

[42] Zamperlini G C M, Santiago-Silva M, Vilegas W (1997). Identification of polycyclic aromatic hydrocarbons in sugar-cane soot by gas chromatography mass spectrometry. *Chromatographia*, 46, 655-663.

[43] Zamperlini G C M, Santiago-Silva M, Vilegas W (2000). Solid-phase extraction of sugar-cane soot extract for analysis by gas chromatography with flame ionisation and mass spectrometric detection. *J. Chromatog.*, A 889, 281-286.

[44] De Andrade S J, Cristale J, Silva F S, Zocolo G J, Marchi M R R (2010). Contribution of sugar-cane harvesting season to atmospheric contamination by polycyclic aromatic hydrocarbons (PAHs) in Araraquara city, Southeast Brazil. *Atmos. Environ.*, 44, 2913-2919.

[45] Cançado J E D, Saldiva P H N, Pereira L A A, Lara L B L S, Artaxo P, Martinelli LA, Arbex MA, Zanobetti A, Braga A L F (2006). The impact of sugar cane-burning emissions on the respiratory system of children and the elderly. *Environ. Hlth. Persp.*, 114, 725-729.

[46] Miranda R M, Andrade M D, Worobiec A, Van Grieken R (2002). Characterisation of aerosol particles in the São Paulo Metropolitan Area. *Atmos. Environ.*, 36, 345-352.

[47] Machado C M D, Cardoso A A, Allen A G (2008). Atmospheric emission of reactive nitrogen during biofuel ethanol production. *Environ. Sci. Technol.*, 42, 381-385.

[48] Gaffney J S, Marley N A (2009). The impacts of combustion emissions on air quality and climate – From coal to biofuels and beyond. *Atmos. Environ.*, 43, 23-36.

[49] Martins E M, Arbilla G (2003). Computer modeling study of ethanol and aldehyde reactivities in Rio de Janeiro urban air. *Atmos. Environ.*, 37, 1715-1722.

[50] Braunbeck O, Bauen A, Rosillo-Calle F, Cortez L (1999). Prospects for green cane harvesting and cane residue use in Brazil. *Biomass Bioenergy*, 17, 495-506.

[51] Castanho D A, Artaxo P (2001). Wintertime and summertime São Paulo aerosol source apportionment study. *Atmos. Environ.*, 35, 4889-4902.

[52] Alonso C D, Martins M H R B, Romano J, Godinho R (1997). São Paulo aerosol characterization study. *J. Air Waste Mgt. Assoc.*, 47, 1297-1300.

[53] Sanchez-Ccoyllo O R, Andrade M D (2002). The influence of meteorological conditions on the behavior of pollutants concentrations in São Paulo, Brazil. *Environ. Poll.*, 116, 257-263.

[54] Miranda R M, Andrade M F (2005). Physicochemical characteristics of atmospheric aerosol during winter in the São Paulo Metropolitan area in Brazil. *Atmos. Environ.*, 39, 6188-6193.

[55] Ynoue R Y, Andrade M D (2004). Size-resolved mass balance of aerosol particles over the São Paulo metropolitan area of Brazil. *Aerosol Sci. Technol.*, 38, 52-62.

[56] Bourotte C, Curl-Amarante A P, Forti M C, Pereira L A A, Braga A L, Lotufo P A (2007). Association between ionic composition of fine and coarse aerosol soluble fraction and peak expiratory flow of asthmatic patients in São Paulo city (Brazil). *Atmos.Environ.*, 41, 2036-2048.

[57] Albuquerque A T T, Andrade M F, Ynoue R Y (2012). Characterization of atmospheric aerosols in the city of São Paulo, Brazil: Comparisons between polluted and unpolluted periods. *Environ. Monit. Assess.*, 184, 969-984.

[58] Gioda A, Sales J A, Cavalcanti P M S, Maia M F, Maia L F P G, Aquino Neto F R (2004). Evaluation of air quality in Volta Redonda, the main metallurgical industrial city in Brazil. *J. Braz. Chem. Soc.*, 14, 856-864.

[59] Toledo V E, Almeida Júnior P B, Quiterio S L, Arbilla G, Moreira A, Escaleira V, Moreira J C (2008). Evaluation of levels, sources and distribution of toxic elements in PM_{10} in a suburban industrial region of Rio de Janeiro, Brazil. *Environ. Monit. Assess.*, 139, 49-59.

[60] Gioia S M C L, Babinski M, Weiss D J, Kerr A A F S (2010). Insights into the dynamics and sources of atmospheric lead and particulate matter in São Paulo, Brazil, from high temporal resolution sampling. *Atmos. Res.*, 98. 478-485.

[61] Miranda R, Tornaz E (2008). Characterization of urban aerosol in Campinas, São Paulo, Brazil. *Atmos. Res.*, 87, 147-157.

[62] Sanchez-Ccoyllo O R, Silva Dias P L, Andrade M D, Freitas S R (2006). Determination of O_3, CO, and PM_{10} transport in the metropolitan area of São Paulo, Brazil through synoptic-scale analysis of back trajectories. *Meteorol. Atmos. Phys.*, 92, 83-93.

[63] Allen A G, Miguel A H (1995). Indoor organic and inorganic pollutants - In-situ formation and dry deposition in southeastern Brazil. *Atmos. Environ.*, 29, 3519-3526.

[64] Vasconcellos P C, Souza D Z, Magalhaes D, Da Rocha G O (2011). Seasonal variation of n-alkanes and polycyclic aromatic hydrocarbon concentrations in PM10 samples collected at urban sites of São Paulo State, Brazil. *Water Air Soil Poll.*, 222, 325-336.

[65] Vasconcellos P C, Souza D Z, Avila S G, Araujo M P, Naoto E, Nascimento K H, Cavalcante F S, Dos Santos M, Smichowski P, Behrentz E (2011). Comparative study of the atmospheric chemical composition of three South American cities. *Atmos. Environ.*, 45, 5770-5777.

[66] Bourotte C, Forti M C, Taniguchi S, Bicego M C, Lotufo P A (2005). A wintertime study of PAHs in fine and coarse aerosols in São Paulo city, Brazil. *Atmos. Environ.*, 39, 3799-3811.

[67] Allen A G, McGonigle A J S, Cardoso A A, Machado C M D, Davison B, Paterlini W, Da Rocha G O, De Andrade J B (2009). Influence of sources and meteorology on surface concentrations of gases and aerosols in a coastal industrial complex. *J. Braz. Chem. Soc.*, 20, 214-221.

[68] Vautz W, Pahl S, Pilger H, Schilling M, Klockow D (2003). Deposition of trace substances via cloud droplets in the Atlantic rain forest of the Serra do Mar, São Paulo State, SE Brazil. *Atmos. Environ.*, 37, 3277-3287.

[69] Allen A G, Da Rocha G O, Cardoso A A, Paterlini W C, Machado C M D (2008). Atmospheric particulate polycyclic aromatic hydrocarbons from road transport in southeast Brazil. *Transportation Res.*, 13, 483-490.

[70] Bourotte C, Forti M C, Melfi A J, Lucas Y (2006). Morphology and solutes content of atmospheric particles in an urban and a natural area of São Paulo State, Brazil. *Water Air Soil Poll.*, 170, 301-316.

[71] Vasconcellos P C, Souza D Z, Sanchez-Ccoyllo O, Bustillos J O V, Lee H, Santos F C, Nascimento K H, Araujo M P, Saarnio K, Teinila K, Hillamo R (2010). Determination of anthropogenic and biogenic compounds in atmospheric aerosol collected in urban, biomass burning and forest areas in São Paulo, Brazil. *Sci. Tot. Environ.*, 408, 5836-5844.

[72] Landulfo E, Papayannis A, Artaxo P, Castanho A D A, Freitas A Z, De Souza R F, Junior N D V, Jorge M, Sánchez-Ccoyllo O R, Moreira D S (2003). Synergetic measurements of aerosols over São Paulo, Brazil using lidar, sunphotometer and satellite data during the dry season. *Atmos. Chem. Phys*, 3, 1523–1539; doi:10.5194/acp-3-1523-2003.

[73] Da Costa R F (2010). Study of the optical properties of aerosols in the State of São Paulo with the Raman Lidar technique. Master Dissertation, Centro de Lasers e Aplicações, Instituto de Pesquisas Energéticas e Nucleares, Universidade de São Paulo, São Paulo, Brazil, 90p.

[74] Torres A S (2008). Development of a methodology for an independent water vapor Raman Lidar calibration to study water vapor atmospheric profiles, PhD Thesis, Centro de Lasers e Aplicações, Instituto de Pesquisas Energéticas e Nucleares, Universidade de São Paulo, São Paulo, Brazil, 144p.

[75] Lopes F J S (2011). Validation of elastic backscatter lidar data from the CALIPSO satellite using the AERONET sun photometer network. PhD thesis, Centro de Lasers e Aplicações, Instituto de Pesquisas Energéticas e Nucleares, Universidade de São Paulo, São Paulo, Brazil, 169p.

[76] Larroza E G (2011). Caracterização das Nuvens Cirrus na Região Metropolitana de São Paulo (RMSP) com a Técnica de Lidar de Retroespalhamento Elástico. PhD thesis, Centro de Lasers e Aplicações, Instituto de Pesquisas Energéticas e Nucleares, Universidade de São Paulo, São Paulo, Brazil, 118p.

[77] Landulfo E, Held G, De Freitas A Z, Papayannis A, De Souza A F (2007). Results from first lidar measurements in the central State of São Paulo during the TroCCiBras 2004 Campaign with IPEN'S aerosol lidar. *Abstracts, 4th Workshop on Lidar Measurements in Latin America*, Ilhabela, S.P., 17-22 June 2007.
http://www.ipen.br/sitio/LWS_Brasil/p4w_abstracts.htm

[78] Held G, Pommereau J-P, Schumann U (2008). TroCCiBras and its partner projects HIBISCUS and TROCCINOX: The 2004 Field Campaign in the State of São Paulo. *Opt.Pura. Apl.*, 41 (2), 207-216. http://www.sedoptica.es/Menu_Volumenes/pdfs/299.pdf

[79] Landulfo E, Lopes F J S, Mariano G L, Torres A S, Jesus W C, Nakaema W M, Jorge M, Mariani R (2010). Study of the properties of aerosols and the air quality index using a backscatter lidar system and AERONET sunphotometer in the city of São Paulo, Brazil. *J. Air Waste Manage. Assoc.*, 60, 386–392; doi: 10.3155/1047-3289.60.4.386.

[80] Landulfo E, Lopes F J S (2009). Initial approach in biomass burning aerosol transport tracking with Calipso and Modis satellites, sunphotometer and a backscatter lidar system in Brazil. *Proceedings of SPIE — The International Society for Optical Engineering*, Vol. 7479, 747905, 2009; doi: 10.1117/12.829973.

[81] Stohl A (1999). *The FLEXTRA Trajectory Model, Version 3.0: User Guide*. Lehrstuhl für Bioklimatologie und Immissionsforschung, University of Munich, Freising, Germany, 1999, 41p. Available from:
http://www.forst.uni-muenchen.de/LST/METEOR/stohl/flextra.htm

[82] Lopes F J S, Landulfo E, Giannakaki E (2008). One-Year of CALIPSO Measurements Over the City of São Paulo, In: *Reviewed and Revised Papers presented at the 24th International Laser Radar Conference 2008*, Boulder, Colorado, v. 2, 1169-1172.

[83] Landulfo E, Jorge M P, Held G, Guardani R, Steffens J, Pinto S A F, Andre I R, Garcia A G, Lopes F J S, Mariano G L, Da Costa R F, Rodrigues P F (2010). Lidar observation campaign of sugar cane fires and industrial emissions in the State of São Paulo, Brazil. *SPIE Digital Library, Proc. SPIE*, Vol. 7832, 783201 (2010), 8p; doi: 10.1117/12.866078.

[84] Mariano G L, Lopes F J S, Steffens J, Jorge M P P M, Landulfo E, Held G, Pinto S A F (2011). Aerosols monitoring in Rio Claro, Brazil, using LIDAR and air pollution analyzers. *Opt. Pura Apl.*, 44, 55-64.

[85] Held G, Lopes F J S, Bassan J M, Nery J T, Cardoso A A, Gomes A M, Ramires T, Lima B R O, Allen A G, Da Silva L C, Souza M L, De Souza K F, Carvalho L R F, Urban R C, Landulfo E, Decco A M, Campos M L A A, Nassur M E Q, Nogueira R F P (2011). Raman lidar monitors emissions from sugar cane fires in the State of São Paulo: A pilot project integrating radar, sodar, aerosol and gas observations. *Revista Boliviana de Física*, 20, 24-26. [Full paper submitted to *Opt. Pura Apl.* in February 2012].

[86] Held G, Larroza E G, Lopes F J S, Da Costa R F, Landulfo E (2011). Raman lidar and sodar measurements in the State of São Paulo, Brazil. *Proceedings, 2011 NDACC Symposium*, Reunion Island, 07-10 November 2011, 4p.

[87] Lopes F J S, Held G, Nakaema W M, Rodrigues P F, Bassan J M, Landulfo E (2011). Initial analysis from a lidar observation campaign of sugar cane fires in the central and western portion of the São Paulo State, Brazil In: Lidar Technologies, Techniques, and Measurements for Atmospheric Remote Sensing VII, 2011. Czech Republic: *Proceedings of SPIE - The International Society for Optical Engineering*, vol. 8182, 818214; doi: 10.1117/12.898119.

[88] Steffens J, Guardani R, Landulfo E, Lopes F J S, Moreira P F, Moreira A (2011). Capability of atmospheric air monitoring in the urban area of Cubatão using lidar technique. *Opt. Pura Apl.*, 44, 65-70.

[89] Dixon M, Wiener G (1993). TITAN: Thunderstorm Identification, Tracking, Analysis & Nowcasting - A radar-based methodology, *J. Atmos. Oceanic Technol.*, 10, 785-797.

[90] Cattrall C, Reagan J, Thome K, Dubovik O (2005). Variability of aerosol and spectral lidar and backscatter and extinction ratios of key aerosol types derived from selected Aerosol Robotic Network locations, *Journal of Geophysical Research*, 110, D10S11. doi:10.1029/2004JD005124.

[91] Omar A H, Winker D M, Kittaka C, Vaughan M A, Liu Z, Hu Y, Trepte C R, Rogers R R, Ferrare R A, Lee K P, Kuehn R E, Hostetler C A (2009). The CALIPSO Automated Aerosol Classification and Lidar Ratio Selection Algorithm, *Journal of Atmospheric and Oceanic Technology*, 26, 1994-2014; doi:10.1175/2009JTECHA1231.1.

[92] Da Silva L C, Allen A G, Cardoso A A, Held G (2011). Aerosol physical and chemical characteristics in the region of Ourinhos (São Paulo State). *Proceedings, Second Conference of the Brazilian Association for Aerosol Research*, Rio de Janeiro, 1-5 August 2011.

[93] De Oliveira B F A, Ignotti E, Hacon S S (2011). A systematic review of the physical and chemical characteristics of pollutants from biomass burning and combustion of fossil fuels and health effects in Brazil. *Cadernos de Saúde Pública*, 27, 1678-1698.

[94] Castro H A D, Cunha M F D, Mendonça G A S, Junger W L, Cunha-Cruz J, Ponce de Leon A (2009). Effect of air pollution on lung function in schoolchildren in Rio de Janeiro, Brazil. *Rev. Saúde Pública*, 43, 26-34.

[95] CETESB (2008). Companhia de Tecnologia de Saneamento Ambiental. Relatório de qualidade do ar no estado de São Paulo 2007. CETESB, São Paulo. http://www.cetesb.sp.gov.br/Ar/publicacoes.asp

[96] Gouveia N, Freitas C U, Martins L C, Marcilio I O (2006). Hospitalizações por causas respiratórias e cardiovasculares associadas à contaminação atmosférica no Município de São Paulo, *Brasil. Cad. Saúde Pública*, 22, 2669-2677.

[97] Braga A L, Saldiva P H, Pereira L A, Menezes J J, Conceição G M, Lin C A, Zanobetti A, Schwartz J, Dockery D W (2001). Health effects of air pollution exposure on children and adolescents in São Paulo, Brazil. *Pediatr. Pulmonol.*, 31, 106-13.

[98] Conceição G M S, Miraglia S G E K, Kishi H S, Saldiva P H N, Singer J M (2001). Air pollution and child mortality: a time-series study in São Paulo, *Brazil. Environ. Hlth. Persp.*, 109, 347-350.

[99] Gouveia N, Fletcher T (2000). Respiratory disease in children and outdoor air pollution in São Paulo, Brazil: a time-series analysis. *Occup. Environ. Med.*, 57, 477-483.

[100] Martins L C, Latorre M R D O, Saldiva P H N, Braga A L F (2001). Relação entre poluição atmosférica e atendimentos por infecção das vias aéreas superiores no

município de São Paulo: avaliação do rodízio de veículos. *Rev. Brás. Epidemiol.*, 4, 220-229.

[101] Martins L C, Latorre M R D O, Saldiva P H, Braga A L (2002). Air pollution and emergency room visits due to chronic lower respiratory disease in the elderly: an ecological times-series study in São Paulo, Brazil. *J. Occup. Environ. Med.*, 44, 622-627.

[102] Miraglia S G E K, Saldiva P H N, Böhm G M (2005). An evaluation of air pollution health impacts and costs in São Paulo, Brazil. *Environ. Mgt.*, 35, 667-676.

[103] Gouveia N, Mendonça G A S, Ponce de Leon A, Correia J E M, Junger W L, Freitas C U, Daumas R P, Martins L C, Guissepe L, Conceição G M S, Manerich A, Cunha-Cruz J (2003). Poluição do ar e efeitos na saúde nas populações de duas grandes metrópoles brasileiras. *Epidemiol. Serv. Saúde*, 12, 29-40.

[104] Freitas C, Bremner S A, Gouveia N, Pereira L A, Saldiva P H N (2004). Internações e óbitos e sua relação com a poluição atmosférica em São Paulo, 1993 a 1997. *Rev. Saúde Pública*, 38, 751-757.

[105] Martins M C H, Fatigati F L, Véspoli T C, Martins L C, Pereira L A A, Martins M A, Saldiva P H N, Braga A L F (2004). Influence of socioeconomic conditions on air pollution adverse health effects in elderly people: an analysis of six regions in São Paulo, Brazil. *J. Epidemiol. Community Health*, 58, 41-46.

[106] Spektor D M, Hofmeister V A, Artaxo P, Brague J A P, Echelar F, Nogueira D P, Hayes C, Thurston G D, Lippmann M (1991). Effects of heavy industrial pollution on respiratory function in the children of Cubatão, Brazil - A preliminary report. *Environ. Hlth. Persp.*, 94, 51-54.

[107] De Martinis B S, Kado N Y, Carvalho L R F, Okamoto R A, Gundel L A (1999). Genotoxicity of fractionated organic material in airborne particles from São Paulo, Brazil. *Mutation Res.*, 446, 83-94.

[108] De Martinis B S, Okamoto R A, Kado N Y, Gundel L A, Carvalho L R F (2002). Polycyclic aromatic hydrocarbons in a bioassay-fractionated extract of PM_{10} collected in São Paulo, Brazil. *Atmos. Environ.*, 36, 307-314.

[109] Vasconcellos P C, Sanchez-Ccoyllo O, Balducci C, Mabilia R, Cecinato A, (2008). Occurrence and concentration levels of nitro-PAH in the air of three Brazilian cities experiencing different emission impacts. *Water Air Soil Poll.*, 190, 87-94.

[110] Vendrasco E P, Silva Dias P L, Freitas E D (2009). A case study of the direct radiative effect of biomass burning aerosols on precipitation in the Eastern Amazon. *Atmos. Res.*, 94, 409-421.

[111] Martins J A, Silva Dias M A F (2009). The impact of smoke from forest fires on the spectral dispersion of cloud droplet size distributions in the Amazonian region. *Env. Res. Lett.*, 4, art. no. 015002.

[112] Moreira J R, Goldemberg J (1999). The Alcohol Program. *Energy Policy*, 27, 229-245.

[113] Arbex M A, Böhm G M, Saldiva P H N, Conceição G M S, Pope III A C, Braga A L F (2000). Assessment of the effects of sugar cane plantation burning on daily counts of inhalation therapy. *J. Air Waste Mgt. Assoc.*, 50, 1745-1749.

[114] Arbex M A, Martins L C, Oliveira R C, Pereira L A, Arbex F F, Cançado J E, Saldiva P H, Braga A L (2007). Air pollution from biomass burning and asthma hospital

admissions in a sugarcane plantation area in Brazil. *J. Epidemiol. Comm. Hlth.*, 61, 395-400.

[115] Lopes F S, Ribeiro H (2006). Mapeamento de internações hospitalares por problemas respiratórios e possíveis associações à exposição humana aos produtos da queima de palha de cana-de-açúcar no estado de São Paulo. Rev. *Brás. Epidemiol.*, 9, 215-225.

[116] Mazzoli-Rocha F, Magalhães C B, Malm O, Saldiva P H, Zin W A, Faffe D S (2008). Comparative respiratory toxicity of particles produced by traffic and sugarcane burning. *Environ. Res.*, 108, 35-41.

[117] De Andrade S J, Varella S D, Pereira G T, Zocolo G J, De Marchi M R R, Varanda E A (2011). Mutagenic activity of airborne particulate matter (PM_{10}) in a sugarcane farming area (Araraquara city, southeast Brazil). *Environ. Res.*, 111, 545-550.

[118] Bell T L, Rosenfeld D, Kim K M, Yoo J M, Lee M I, Hahnenberger M (2008). Midweek increase in US summer rain and storm heights suggests air pollution invigorates rainstorms. *J. Geophys. Res.*, 113, D2, Art. No. D02209.

[119] IPCC (2007). Fourth Assessment Report "Climate Change 2007", International Panel on Climate Change: www.ipcc.ch.

[120] Rosenfeld D (2006). Aerosol-cloud interactions control of earth radiation and latent heat release budgets. *Space Sci. Rev.*, 125, 149-157.

[121] Dufek A S, Ambrizzi T (2008). Precipitation variability in São Paulo State, Brazil. *Theoret. Appl. Climatol.*, 93, 167-178.

[122] Coelho C H, Allen A G, Fornaro A, Orlando E A, Grigoletto T L B, Campos M L A M (2011). Wet deposition of major ions in a rural area impacted by biomass burning emissions. *Atmos. Environ.*, 45, 5260-5265.

[123] UNICA (2012). http://english.unica.com.br/dadosCotacao/estatistica/ (Accessed 29/4/2012).

[124] Gomes A M, Escobedo J F (2010). Climatologia de Tempestades na Área Central do Estado de São Paulo Usando Radar Meteorológico. *Revista Energia na Agricultura*, 25, no. 1, 1-20. ISSN- 1808-875.

[125] Westcott N E (1995). Summertime cloud-to-ground lightning activity around major Midwestern urban areas. *J. Appl. Meteor.*, 34, 1633-1642.

[126] UNICA (2011). (Accessed 17/4/2012) http://english.unica.com.br/website/userFiles/protocolo-agroambiental.pdf

[127] Tsao C C, Campbell J E, Mena-Carrasco M, Spak S N, Carmichael G R, Chen Y (2012). Increased estimates of air-pollution emissions from Brazilian sugar-cane ethanol. *Nature Climate Change*, 2, 53-57.

Chapter 2

The Chemistry of Dicarboxylic Acids in the Atmospheric Aerosols

Mohd Zul Helmi Rozaini

Additional information is available at the end of the chapter

1. Introduction

Atmospheric chemistry is a branch of atmospheric science in which the chemistry of the Earth's atmosphere and that of other planets is studied. It is a multidisciplinary field of research and draws on environmental chemistry, physics, meteorology, computer modeling, oceanography, geology and volcanology and other disciplines. It also deals with chemical compounds in the atmosphere, their distribution, origin, chemical transformation into other compounds and finally their removal from the atmospheric domain. These substances may occur as gasses, liquids or solid. The composition of the atmosphere is dominated by the gasses nitrogen and oxygen in proportions that have been found to be invariable in time and space at altitudes up to 100 km. All other compounds are minor ones, with many of them occurring only in traces.

The composition and chemistry of the atmosphere is of importance for several reasons, but primarily because of the interactions between the atmosphere and living organisms. The composition of the Earth's atmosphere (Figure 1) has been changed by human activity and some of these changes are harmful to human health, crops and ecosystems. Examples of problems which have been addressed by atmospheric chemistry include acid rain, photochemical smog and global warming. Atmospheric chemistry seeks to understand the causes of these problems, and by obtaining a theoretical understanding of them, allow possible solutions to be tested and the effects of changes in government policy evaluated.

Observations, lab measurements and modeling are the three important methodologies in atmospheric chemistry. Progress in atmospheric chemistry is often driven by the interactions between these components and they form an integrated whole. For example observations may tell us that more of a chemical compound exists than previously thought possible. This will stimulate new modelling and laboratory studies which will increase our scientific understanding to a point where the observations can be explained. Measurements made in the laboratory are essential to our understanding of the sources and sinks of pollutants and naturally occurring compounds. Lab studies tell us which gases react with each other and how fast they react. Measurements of interest include reactions in the gas phase, on surfaces and in water. Also of high importance is photochemistry which quantifies how quickly molecules are split apart by sunlight and what the products are plus thermodynamic data such as Henry's law coefficients.

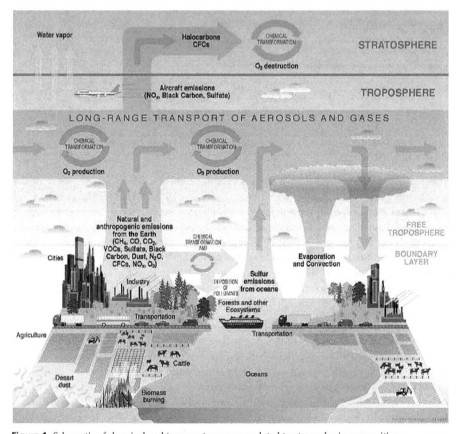

Figure 1. Schematic of chemical and transport processes related to atmospheric composition.

Modelling for instance is important to synthesize and test theoretical understanding of atmospheric chemistry. Computer models (such as chemical transport models) are used. Numerical models solve the differential equations governing the concentrations of chemicals in the atmosphere. They can be very simple or very complicated. One common trade off in numerical models is between the number of chemical compounds and chemical reactions modelled versus the representation of transport and mixing in the atmosphere. For example, a box model might include hundreds or even thousands of chemical reactions but will only have a very crude representation of mixing in the atmosphere. In contrast, 3D models represent many of the physical processes of the atmosphere but due to constraints on computer resources will have far fewer chemical reactions and compounds. Models can be used to interpret observations, test understanding of chemical reactions and predict future concentrations of chemical compounds in the atmosphere. One important current trend is for atmospheric chemistry modules to become one part of earth system models in which the links between climate, atmospheric composition and the biosphere can be studied.

2. Background knowledge

2.1. Aerosol

An aerosol is a system (in the sense of a system as used in thermodynamics or chemistry) comprising liquid and/or solid particles in a carrier gas. It is generally defined as a suspension of liquid or solid particles in a gas, with particle diameters in the range of 10^{-9}-10^{-4} m (lower limit: molecules and molecular clusters: upper limit: rapid sedimentation). The most evident examples of aerosols in the atmosphere are clouds, which consist primarily of condensed water. The suspension of the particles in the gas must be significantly stable and homogenous. Hence the assumptions of stability and homogeneity, and consequently the possibilities to use statistical descriptors, are limited to understand and to predict the system, the particle properties, i.e. their size, shapes, chemical compositions, their surfaces, their optical properties, their volumes and masses must be known (Preining, 1993). Aerosol particles scatter and absorb solar and terrestrial's radiation, they are involved in the formation of clouds and precipitation as cloud condensation and ice nuclei, and they affect the abundance and distribution of atmospheric traces gases by heterogeneous chemical reactions and other multiphase processes.

2.2. Aerosol types

The atmospheric aerosol in the boundary layer and the lower troposphere is different for different regions, the main types are:

a. continental aerosol - a main component of which is mineral dust;
b. maritime aerosol - a main component of which is sea salt;
c. background aerosol - aged accumulation mode aerosol.

Chemically or photochemically produced from precursor gases, continental or oceanic biosphere or from anthropogenic releases including sulphates, nitrates, hydrocarbons, soot and so on. The continental aerosols are strongly influenced by man's activities and include urban and rural aerosols. Dust storms produce another type of continental aerosol. Aerosols with a lifetime of up to several years exist in the stratosphere, the sources of which are volcanic injections, and particles or gases entering the stratosphere via diffusion from the troposphere as well as interplanetary dust entering from space. The most important source is volcanic injection. Due to their long lifetime, these aerosols are distributed relatively homogeneously throughout the whole stratosphere and the size distribution is unimodal with only the accumulation mode present.

2.3. The study of atmospheric aerosols

Atmospheric aerosol particles are a ubiquitous part of earth's atmosphere, present in very lungful of air breathed. They are produced in vast numbers by both human activity (anthropogenic) and natural sources and subsequently modified by a multitude processes. They are known to be crucially important in many issues that directly affect everyday life which include respiratory health, visibility, clouds, rainfall, atmospheric chemistry and global regional climate but they are also one of the more poorly understood aspects of the atmosphere. These shortcomings in understanding are partly due to their small size, which is typically of the order of microns or less, making them difficult to study and also the fact that the processes involved are complex. The description of the organic chemistry in atmospheric aerosol is by no means straightforward, but the addition of the solubility variables, aerosol thermodynamic, hygroscopic properties, deliquescence behaviour makes understanding the atmosphere and its effect is even more challenging, requiring the application of wide spectrum of scientific disciplines including chemistry, physics, mechanics, biology and medicine.

2.4. Aerosols and effect on quality of life

The effects of aerosols on the atmosphere, climate and public health are among the central topics in current environmental research. Urban areas have always been known to be a major source of particulate pollution (Finlayson-Pitss, 2000) which is expected to continue to increase due to world population growth and increasing industrialization and energy use, especially in developing countries (Fenger, 1999). The most obvious effects are the contributions to unsightly smogs and visible deterioration of the building materials (Grossi, 2002). In addition, the fact that urban particulate pollution impact directly on human health has been known for centuries (Brimblecombe, 1987) and has been the subject of much research (Adam et al., 1999).

In an attempt to reduce the health burden of atmospheric particulate pollution, regulatory authorities have attempted to place controls on the emission and the magnitude of pollution episodes within conurbations. The monitoring of particulate air pollution has traditionally focused on particles of less than 10 μm in aerodynamic diameter (the PM_{10} standard), as these are more likely to pass the throat when inhaled (DEFRA, 2005; Larrsen, 1999) but it has become apparent that the smaller particles are more significant, as these particles will penetrate deeper into the lungs and potentially cause more physiological distress or damage. This has lead to the use of the $PM_{2.5}$ standard in countries such Malaysia, where the total mass of particulate matter less than 2.5 μm in diameter is monitored (MOSTI, 2000).

2.5. Composition of atmospheric aerosol

The atmospheric aerosol consists of a complex mixture of organic and inorganic compounds (Cruz, 1998). The typical composition of fine continental aerosol will usually contain various sulphates (mostly ammonium and calcium), nitrates (mostly ammonium), chlorides (mostly sodium), elemental carbon (EC) and organic carbon (OC), especially traffic-related soot, biological materials and other organic compounds, iron compounds, trace metals, and mineral derived from rocks, soil and various human activities. Aerosol composition also can be influenced by local geology, geographic location and climate (Moreno et al., 2003).

2.5.1. Organic and elemental carbon of aerosol

Several studies have shown that over 30% of aerosol is organic carbon, and carbon containing matter can account for as much as 50%. Typically, two classes of carbonaceous aerosol are commonly present in ambient air: organic carbon (OC) and elemental carbon (EC), which are the largest contributors to the fine particle burden in urban atmospheres and heavily industrialised areas (Cachier et al., 1989).

Field measurements also shown a significant mass fraction of atmospheric aerosol consist of organic compounds (Rogge et al., 1991). Around 5 to 10% of the known fraction is often limited to low molecular weight species, which are identified by standard analytical techniques, using gas chromatography coupled with mass spectrometry. A significant fraction of the organic mass in tropospheric aerosol, is comprised of high molecular weight, oxygenated species which remain unidentified (Decesari et al., 2002).

Organic compounds are emitted into the atmosphere from various anthropopgenic and biogenic sources. These include primary emission, mainly from combustion and biogenic sources and secondary organic aerosol resulting from the reaction of primary volatile organic compounds in the atmosphere (Fisseha et al., 2004). In urban areas, a number of emission sources are responsible for the presence of organic aerosol in the atmosphere among which are road traffic, industrial processes, waste incineration, wastewater treatment processes and domestic heating. Some of these are pure organic aerosols, which may be formed by primary particle emissions (primary organic carbon) or produced from atmospheric reactions involving gaseous organic precursors (secondary OC)(Cruz and Pandis, 1998).

Organic material is important in controlling the aerosol physico-chemical properties (Cornell et al., 2003). They also found that the uptake of liquid water in aerosol was enhanced by the presence of organic carbon compounds. Organic carbon is also an effective light scatter and may contribute significantly to both visibility degradation and direct aerosol climate forcing (Heintzenberg., 1989). Elemental carbon (often named black carbon or soot) may be the second most important elemental in global warming in terms of direct forcing, after CO_2 due to specific surface properties. Elemental carbon provides a good adsorbtion site for many semi-volatile compounds such as poly-aromatic hydrocarbon (PAH) and offers a large specific surface area for interactions with reactive trace gases such as ozone. Annually, about 13 Tg black carbons are emitted into the atmosphere, mainly through fossil fuel combustion and biomass burning (Jacob, 1999).

As for other aerosols, the removal of particulate carbon is likely to occur via two main scavenging processes: the in-cloud process, whereby particles are directly incorporated into cloud droplets; and the below-cloud process, where particles are washed out by precipitation itself. The physico-chemical atmospheric processes which transform young combustion particles, expected to be hydrophobic, into a water soluble aerosol phase remains a major unknown. The atmospheric behaviour of the carbonaceous particles is likely to be dictated by the chemical nature of their surfaces (Cachier et al., 1989). If the surface is hydrophobic, the particle remains inactive. However, if it is coated with hygroscopic substances, it may be activated enough to be incorporated into water droplets (Charlson and Heintzenberg, 1995).

2.5.2. Water soluble organic compounds

A significant fraction of the particulate organic carbon is water soluble, ranging from 20% to 70% of the total soluble mass, thus making it important to various aerosol-cloud interactions (Decesari et al., 2000; Facchini et al., 2000). Water soluble organic compounds (WSOC) contribute to the ability of the particles to act as cloud condensation nuclei (CCN) (Novokov and Penner, 1993).

WSOC have been postulated to be partially responsible for the water uptake of airbone particulate matter, which can substantially affect the physical and chemical properties of atmospheric aerosols (Yu et al., 2005). Decesari et al. (2001) have suggested that WSOC are composed of higly oxidised species with residual aromatic nuclei and aliphatic chains. The current understanding of atmospheric particles describes their WSOC fraction as a complex mixture of very soluble organic compounds, slightly soluble organic compounds, and some undetermined macromolecular compounds (MMCs)(Saxena and Hildemann, 1996).

The composition of WSOC varies among sampling regions. It was found to constitute between 20 and 67% of the total organic carbon present in aerosol samples collected in Tokyo (Sempere and Kawamura, 1994). The percentage is ranged from 65 to 75% in aerosol samples collected in Hungary, Italy and Sweeden (Zappoli et al., 1999). The study also found that the percentage of WSOC species with respect to the total soluble mass was much higher at the background site (Aspvreten, Central Sweeden) (c.a. 50%) compared to the polluted site (San Pietro Copofiume, Po Valley, Italy) (c.a. 25%). A very high fraction (over 70%) of organic compounds in the aerosol consisted of polar species. A study by Wang et al. (2002) showed that most water soluble carbon is total organic carbon (TOC) and range between 20.53 to 35.58 μg m^{-3} in PM$_{10}$ and PM $_{2.5}$. A further study by (Narukawa et al., 1999) concluded that individual haze particles over Kalimantan of Indonesia were mainly composed of water soluble organic materials and inorganic salt such as ammonium sulphate.

The ionic organic compounds (including carboxylic, dicarboxylic and ketoacids) were distributed between both sub-micron and super micron mode, indicating origins in both gas-to-particle conversion and heterogeneous reaction on pre-existing particles. WSOC in atmospheric aerosols and droplets can be divided by their functional groups into three classes which are neutral, mono- and dicarboxylic acid and also polycarboxylic acid, which were found to account on average for 87% of total fine aerosol WSOC (Decesari et al., 2000). The most frequently determined WSOC are the low molecular weight (LMW) carboxylic and dicarboxylic acids (Yu, 2000). Most of carboxylic acids compound are a secondary oxidation products of atmospheric organic compounds and also found in remote marine as well as continental rural and urban areas (Simoneit and Mazurek, 1982). Among these dicarboxylic acids (DCA's), oxalic acid is the most abundant, followed by succinic and malonic in atmospheric aerosol especially during summer season.

In the aqueous phase, organic oxidation also can be initiated by various radical anions in the atmosphere (e.g. OH^-,NO_3^-,SO_4^{2-},Cl^-). Among these species, it is very likely that $OH \cdot$ is the most efficient iniating organic oxidation (Dutot et al., 2003). The DCA's are the late products in the photochemistry of aliphatic and aromatic hydrocarbons, and due to the low vapour pressure, it is almost entirely partitioned to the particulate phase. They also constitute an important fraction of the water soluble part of particulate organic matter (POM) in atmospheric aerosol particles at remote and urban areas (Rohrl and Lammel, 2001).

3. Dicarboxylic acids

During the past decade, much attention has been paid to the low molecular weight dicarboxylic acids and related polar compounds which are ubiquitous water-soluble organic compounds that have been detected in a variety of environmental samples including atmospheric aerosols, rainwaters, snow packs, ice cores, meteorites, marine sediments, hypersaline brines and freshwaters (Kawamura and Ikushima, 1993; Tedetti et al., 2006). In the atmosphere, dicarboxylic acids originate from incomplete combustion of fossil fuels (Kawamura and Ikushima, 1993; Kawamura and Kaplan, 1987), biomass burning (Narukawa et al., 1999), direct biogenic emission and ozonolysis and photo-oxidation of organic compound (Sempere and Kawamura, 2003).

Low molecular weight (LMW) dicarboxylic acids have also been identified in cloud water samples collected at a high mountain range in central europe (Puxbaum and Limbeck, 2000), in the condensed phase at a semi-urban site in the northeastern US (Khwaja, 1995) and in Arctic aerosol (Kawamura et al., 1996). As a result of their hygroscopic properties, dicarboxylic acids can act as cloud condensation nuclei and have an impact on the radiative forcing at earth's surface (Kerminen et al., 2000). Dicarboxylic acids also participate in many biological processes. They are important intermediates in the tricarboxylic acid and glyoxylate cycles and the catabolism and anabolism of amino acids (Tedetti et al., 2006).

Photochemical reactions are also an important source of atmospheric dicarboxylic acids. For example, glutaric acids photooxidation is likely the dominant pathway formation, as measured atmospheric concentrations of dicarboxylic acids in Los Angeles far surpasses contributions from direct emissions and seasonal trends suggest that dicarboxylic acids are largely produced in photochemical smog (Puxbaum and Limbeck, 2000; Rogge et al., 1993).

Aliphatic dicarboxylic acids (or diacids) can be described by the following general formula:

$$HOOC\text{-}(CH_2)_n\text{-}COOH$$

According to IUPAC nomenclature, dicarboxylic acids are named by adding the suffix dioic acid to the name of the hydrocarbon with the same number of carbon atoms, e.g., nonanedioic acid for $n = 7$. The older literature often uses another system based on the hydrocarbon for the $(CH_2)_n$ carbon segment and the suffix dicarboxylic acid, e.g., heptanedicarboxylic acid for $n = 7$. However, trivial names are commonly used for the saturated linear aliphatic dicarboxylic acids from $n = 0$ (oxalic acid) to $n = 8$ (sebacic acid) and for the simple unsaturated aliphatic dicarboxylic acids; these names are generally derived from the natural substance in which the acid occurs or from which it was first isolated.

Aliphatic dicarboxylic acids are found in nature both as free acids and as salts. For example, malonic acid is present in small amounts in sugar beet and in the green parts of the wheat plant; oxalic acid occurs in many plants and in some minerals as the calcium salt. However, natural sources are no longer used to recover these acids.

The main industrial process employed for manufacturing dicarboxylic acids is the ring-opening oxidation of cyclic compounds.

Oxalic acid is the most important dicarboxylic acid. Adipic, malonic, suberic, azelaic, sebacic, and 1,12-dodecanedioic acids, as well as maleic and fumaric acids, are also manufactured on an industrial scale.

Physical properties: Dicarboxylic acids are colorless, odorless crystalline substances at room temperature. Table 1 lists the major physical properties of some saturated aliphatic dicarboxylic acids.

The lower dicarboxylic acids are stronger acids than the corresponding monocarboxylic ones. The first dissociation constant is considerably greater than the second. Density and dissociation constants decrease steadily with increasing chain length. By contrast, melting point and water solubility alternate: Dicarboxylic acids with an even number of carbon atoms have higher melting points than the next higher odd-numbered dicarboxylic acid. In the $n = 0 - 8$ range, dicarboxylic acids with an even number of carbon atoms are slightly soluble in water, while the next higher homologues with an odd number of carbon atoms are more readily soluble. As chain length increases, the influence of the hydrophilic carboxyl groups diminishes; from $n = 5$ (pimelic acid) onward, solubility in water decreases rapidly. The alternating solubility of dicarboxylic acids can be exploited to separate acid mixtures. Most dicarboxylic acids dissolve easily in lower alcohols; at room temperature, the lower dicarboxylic acids are practically insoluble in benzene and other aromatic solvents.

IUPAC name	Common name	Formula	Molecular weight, M_r	Melting point, M_p, °C	Boiling point, B_p at 13.3 kPa,°C	Density, l at 25°C g/cm³	Solubility in mol/kg	Ionisation constant	
								K_1	K_2
Ethanedioc acid	Oxalic acid	HOOC-COOH	90.03	189.5	-	1.653	1.131	5.29×10^{-2}	5.33×10^{-5}
Propanedioc acid	Malonic acid	HOOC-CH2-COOH	104.06	135	-	1.619	16.03	1.42×10^{-3}	2.01×10^{-6}
Butanedioic acid	Succinic acid	HOOC-(CH2)2-COOH	118.08	188	235	1.572	0.748	6.16×10^{-5}	2.31×10^{-6}
Pentanedioic acid	Glutaric acid	HOOC-(CH2)3-COOH	132.11	99	200	1.424	8.468	4.57×10^{-5}	3.89×10^{-6}
Hexanedioic acid	Adipic acid	HOOC-(CH2)4-COOH	146.14	153	265	1.360	0.171	3.85×10^{-5}	3.89×10^{-6}
Heptanedioic acid	Pimelic acid	HOOC-(CH2)5-COOH	160.17	106	272	1.329	0.423	3.19×10^{-5}	3.74×10^{-6}
Octanedioic acid	Suberic acid	HOOC-(CH2)6-COOH	174.19	144	279	1.266	0.0139	3.05×10^{-5}	$3.85 \times^{-6}$
Nonanedioic acid	Azelaic acid	HOOC-(CH2)7-COOH	188.22	108	287	1.225	0.00946	2.88×10^{-5}	3.86×10^{-6}
Decanedioic acid	Sebacic acid	HOOC-(CH2)8-COOH	202.25	134.5	-	1.207	0.00012	3.1×10^{-5}	3.6×10^{-6}

Table 1. Physical properties of saturated dicarboxylic acid (Clarke, 1986)

Chemical properties: The chemical behavior of dicarboxylic acids is determined principally by the two carboxyl groups. The neighboring methylene groups are activated generally to only a minor degree. Thermal decomposition of dicarboxylic acids gives different products depending on the chain length. Acids with an even number of carbon atoms require higher decarboxylation temperatures than the next higher odd-numbered homologues; lower dicarboxylic acids decompose more easily than higher ones. To avoid undesired decomposition reactions, aliphatic dicarboxylic acids should only be distilled in vacuum. When heated above 190 °C, oxalic acid decomposes to carbon monoxide, carbon dioxide, and water. Malonic acid is decarboxylated to acetic acid at temperatures above 150 C:

$$HOOC\text{-}(CH_2)_n\text{-}COOH\text{-}CH_3COOH + CO_2$$

When malonic acid is heated in the presence of P_2O_5 at ca. 150 °C, small amounts of carbon suboxide (C_3O_2) are also formed. Succinic and glutaric acids are converted into cyclic anhydrides on heating:

where n = 2 or 3

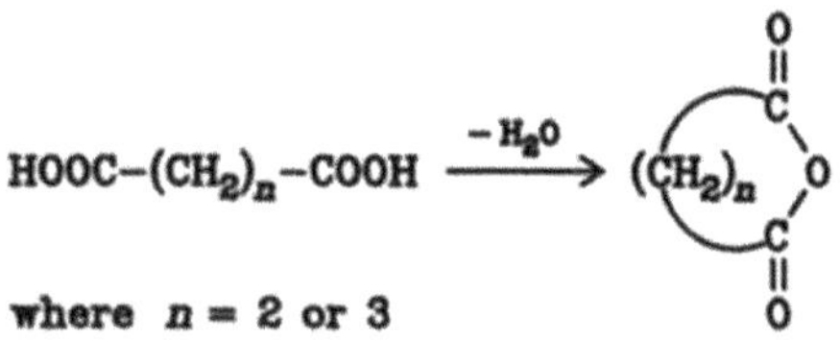

Scheme 1. Succinic and glutaric acids are converted into cyclic anhydrides on heating

When the ammonium salt of succinic acid is distilled rapidly, succinimide is formed, with the release of water and ammonia.

Higher dicarboxylic acids from $n = 4$ (adipic acid) to $n = 6$ (suberic acid) split off carbon dioxide and water to form cyclic ketones:

where n = 4-6

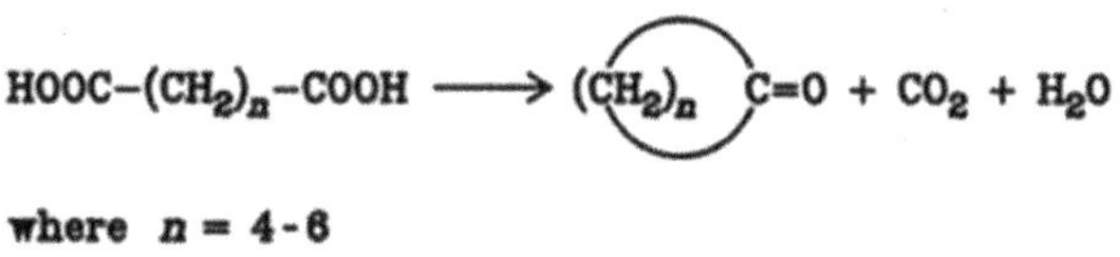

Scheme 2. Higher dicarboxylic acids from n = 4 (adipic acid) to n = 6 (suberic acid) split off carbon dioxide and water to form cyclic ketones

The decomposition of still higher dicarboxylic acids leads to complex mixtures. With the exception of oxalic acid, dicarboxylic acids are resistant to oxidation. Oxalic acid is used as a reducing agent for both commercial and analytical purposes. Dicarboxylic acids react with dialcohols to form polyesters and with diamines to form polyamides. They also serve as starting materials for the production of the corresponding diamines. Reaction with monoalcohols yields esters. All of these reactions are commercially important. Several reactions with malonic and glutaric acids are of interest in organic syntheses: the Knoevenagel condensation, Michael addition, and malonic ester synthesis (Clarke, 1986)

Succinic acid ester reacts with aldehydes or ketones in the presence of sodium ethoxide or potassium *tert*-butoxide to form alkylidenesuccinic acid monoesters (Stobbe condensation). These can subsequently be converted into monocarboxylic acids by hydrolysis, decarboxylation, and hydrogenation (Clarke, 1986).

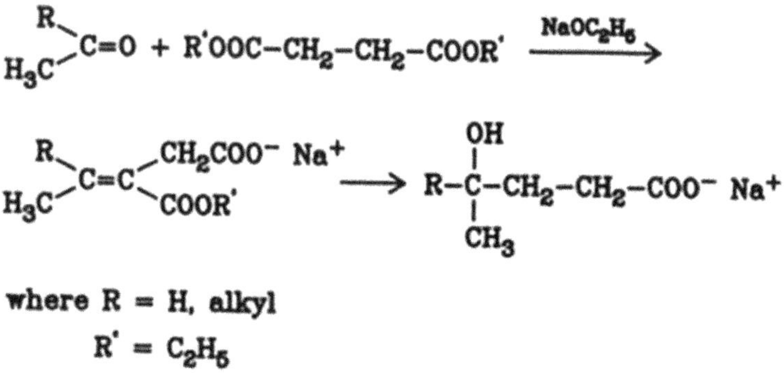

$$\underset{H_3C}{\overset{R}{\diagdown}}C{=}O \ + \ R'OOC{-}CH_2{-}CH_2{-}COOR' \ \xrightarrow{\ NaOC_2H_5\ }$$

$$\underset{H_3C}{\overset{R}{\diagdown}}C{=}C\underset{COOR'}{\overset{CH_2COO^- \ Na^+}{\diagup}} \ \longrightarrow \ R{-}\underset{\underset{CH_3}{|}}{\overset{\overset{OH}{|}}{C}}{-}CH_2{-}CH_2{-}COO^- \ Na^+$$

where R = H, alkyl

R' = C₂H₅

Scheme 3. Production number of straight-chain aliphatic dicarboxylic acids and their derivatives occur in nature

Production: A number of straight-chain aliphatic dicarboxylic acids and their derivatives occur in nature. However, isolation from natural substances has no commercial significance. Although many syntheses for the production of aliphatic dicarboxylic acids are known, only a few have found industrial application. This is due partly to the shortage of raw materials.

Individual saturated dicarboxylic acids: Dicarboxylic acids are used mainly as intermediates in the manufacture of esters and polyamides. Esters derived from monofunctional alcohols serve as plasticizers or lubricants. Polyesters are obtained by reaction with dialcohols. In addition, dicarboxylic acids are employed in the manufacture of hydraulic fluids, agricultural chemicals, pharmaceuticals, dyes, complexing agents for heavy-metal salts, and lubricant additives (as metal salts).

3.1. Oxalic acid

Oxalic acid (ethanedioic acid, acidum oxalicum) is the simplest saturated dicarboxylic acid (Clarke, 1986). The compound exists in anhydrous form [144-62-7] or as a dihydrate [6153-56-6]. The anhydrous acid is not found in nature and must be prepared from the dihydrate even when produced industrially. Oxalic acid is widely distributed in the plant and animal kingdom (nearly always in the form of its salts) and has various industrial applications.

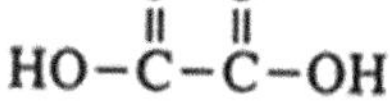

$$HO{-}\underset{}{\overset{\overset{O}{\|}}{C}}{-}\underset{}{\overset{\overset{O}{\|}}{C}}{-}OH$$

Scheme 4. Chemical structure of oxalic acid

The acidic potassium salt of oxalic acid is found in common sorrel (Latin: oxalis acetosella) and the name oxalic acid is derived from that plant. Table 2 shows examples of plants in which oxalic acid occurs (in the form of potassium, sodium, calcium, magnesium salts, or iron complex salts) are given below (oxalic acid content in milligrams per 100 g dry weight):(Tsu-Ning Tsao G., 1963)

Spinach	460 – 3200
Rhubarb	500 – 2400
Chard	690
Parsley	190
Beets	340
Cocoa	4500
Tea	3700
Beet leaves	up to 12 000

Table 2. Oxalic acid content in milligrams per 100 g dry weight

Oxalic acid is formed in plants through incomplete oxidation of carbohydrates, e.g., by fungi (*Aspergillus niger*) or bacteria (*acetobacter*) and in the animal kingdom through carbohydrate metabolism via the tricarboxylic acid cycle. The urine of humans and of most mammals also contains a small amount of calcium oxalate. In pathological cases, an increased calcium oxalate content in urine leads to the formation of kidney stones (Clarke, 1986). Calcium and iron(II) oxalates are also found as minerals. Both the anhydrous and dihydrated forms of oxalic acid form colorless and odorless crystals.

Anhydrous oxalic acid

Anhydrous oxalic acid [144-62-7] exists as rhombic crystals in the *a*-form and as monoclinic crystals in the *b*-form (West, 1980). These forms differ mainly in their melting points. The slightly stable *b*-form changes into the *a*-form at 97 °C and 0.2 barr. Anhydrous oxalic acid is prepared by dehydration of the dihydrate through careful heating to 100 °C. It is then sublimated in a dry air stream. The sublimation is fast at 125 °C and can be carried out at temperatures up to 157 °C without decomposition. The dehydration can also be accomplished by azeotropic distillation with benzene or toluene. Anhydrous oxalic acid is slightly hygroscopic; it absorbs water from moist air ("weathers") to form the dihydrate again. The hydration occurs very slowly because of surface caking.

Oxalic acid dihydrate

Oxalic acid dihydrate [6153-56-6], HOOC–COOH · 2 H₂O is the industrially produced and usual commercial form of oxalic acid. The compound forms colorless and odorless prisms or granules that contain 71.42 wt % oxalic acid and 28.58 wt % water. Oxalic acid dihydrate is stable at room temperature and under normal storage conditions. The most important physical properties are as follows:

The solubility in water and the density of these solutions are presented in Table.1. Oxalic acid is readily soluble in polar solvents such as alcohols (although partial esterification occurs), acetone, dioxane, tetrahydrofuran, and furfural. Oxalic acid is sparingly soluble in diethyl ether (1.5 g oxalic acid dihydrate in 100 g ether at 25 °C), and insoluble in benzene, chloroform, and petroleum ether. The ionization constants show that oxalic acid is a strong acid. The value of K_1 is comparable to that of mineral acids and the value of K_2 corresponds to ionization constants of strong organic acids, for example, benzoic acid.

In the homologous series of dicarboxylic acids, oxalic acid, the first member, shows unique behavior because of the interaction of the neighboring carboxylate groups. This results in an increase in the value of the dissociation constant and in the ease of decarboxylation: Upon rapid heating to 100 °C oxalic acid decomposes into carbon monoxide, carbon dioxide, and water with formic acid as an isolable intermediate.

In aqueous solution decomposition is induced by light and to a much greater extent by g- or X-rays (to carbon monoxide, carbon dioxide, formic acid, and occasionally hydrogen). This decomposition is catalyzed by the salts of heavy metals, for example, by uranyl salts. Oxalic acid cannot form an intramolecular anhydride. Upon heating to over 190 °C or warming in concentrated sulfuric or phosphoric acid, oxalic acid decomposes to carbon monoxide, carbon dioxide, and water: this decomposition is not exothermic.

The reducing properties of oxalic acid (which itself is oxidized to the harmless end products carbon dioxide and water) form the basis for the variety of practical applications. Oxalic acid is also oxidized relatively easily to carbon dioxide by many other oxidizing agents in addition to air, especially in the presence of the salts of heavy metals. Oxalic acid is easily esterified, whereby two types, the acidic mono or neutral diesters can result. These esters are applied as intermediates in chemical syntheses. They react relatively easily with water, ammonia, or amines to afford the corresponding acyl derivatives.

Important chemical characteristics are also demonstrated by the metal salts of oxalic acid. These exist in two types-the acidic and neutral salts. The alkali metal and iron (III) salts are readily soluble in water. All other salts are sparingly soluble in water. The near complete insolubility of the alkaline-earth salts of oxalic acid, especially of calcium oxalate, finds some applications in quantitative analysis. When heated all these metal salts lose carbon monoxide. Other salts which are easier decomposable lose carbon dioxide in addition. The alkali and alkaline-earth salts form carbonates under these conditions. Manganese, zinc, and tin salts form oxides; iron, cadmium, mercury, and copper salts form mixtures of oxides and metals. Nickel, cobalt, and silver salts afford pure metals. Anhydrous fusion of oxalates with alkali yield carbonates and hydrogen. For a review see Dollimore (1987).

3.2. Malonic acid

Three-carbon 1,3-dicarboxylic acid derivatives (malonic acid, malonates, cyanoacetic acid, cyanoacetates, and malononitrile) are widely used in industry for the manufacture of pharmaceuticals, agrochemicals, vitamins, dyes, adhesives, and fragrances. The common feature of malonic acid and its derivatives is the high reactivity of the central methylene group. Due to the increasingly electron-withdrawing character of the substituents, the acidity of the hydrogen atoms in the 2-position increases in the order malonates < cyanoacetates < malononitrile. Therefore, all these compounds undergo reactions typical of 1,3-dicarbonyl compounds. For example they are easily alkylated or arylated, undergo aldol and Knoevenagel condensations, and they can be used for the synthesis of pyrimidines and other nitrogen heterocycles.

Physical Properties: Important physical properties of malonic acid (propanedioic acid, methanedicarboxylic acid) are listed in Table 1. Its pK_a values are 2.83 and 5.70. Malonic acid forms a colorless hygroscopic solid which sublimes in vacuum with some decomposition. It's really soluble in the water; but slightly soluble in ethanol and diethyl ether, and is completely insoluble in benzene.

Chemical Properties: Malonic acid is found in small amounts in sugar beet and green wheat, being formed by oxidative degradation of malic acid. Reaction with sulfuryl chloride or bromine gives mono- and dihalogenated malonic acid, whereas treatment with thionyl chloride or phosphorus pentachloride leads to mono- or diacyl chloride. When heated with phosphorus pentoxide, malonic acid does not form an anhydride but rather carbon suboxide, a toxic gas that reacts violently with water to reform malonic acid. On heating the free acid above 130 °C, or an aqueous solution above 70 °C, decomposition to acetic acid and carbon dioxide takes place. The mono- and dianion of malonic acid are more stable. In aqueous solution the monosodium salt decomposes above 90 °C and the disodium salt above 130 °C (Bolton, 1995).

3.3. Succinic acid

Succinic acid is found in amber, in numerous plants (e.g., algae, lichens, rhubarb, and tomatoes), and in many lignites.

Production: A large number of syntheses are used to manufacture succinic acid. Hydrogenation of maleic acid, maleic anhydride, or fumaric acid produces good yields of succinic acid; the standard catalysts are Raney nickel, Cu, NiO, or CuZnCr, Pd – Al$_2$O$_3$, Pd – CaCO$_3$, or Ni – diatomite. 1,4-Butanediol can be oxidized to succinic acid in several ways: (1) with O$_2$ in an aqueous solution of an alkaline-earth hydroxide at 90 – 110 °C in the presence of Pd – C; (2) by ozonolysis in aqueous acetic acid; or (3) by reaction with N$_2$O$_4$ at low temperature. Succinic acid or its esters are also obtained by Reppe carbonylation of ethylene glycol, catalyzed with RhCl$_3$ – pentachlorothiophenol; Pd-catalyzed methoxycarbonylation of ethylene; and carbonylation of acetylene, acrylic acid, dioxane, or β- propiolactone (Bolton, 1995).

Acid mixtures containing succinic acid are obtained in various oxidation processes. Examples include the manufacture of adipic acid; the oxidation of enanthic acid and the ozonolysis of palmitic acid. Succinic acid can also be obtained by phase-transfer-catalyzed reaction of 2-haloacetates, electrolytic dimerization of bromoacetic acid or ester, oxidation of 3-cyanopropanal, and fermentation of *n*-alkanes.

Uses*: Succinic acid is used as a starting material in the manufacture of alkyd resins, dyes, pharmaceuticals, and pesticides. Reaction with glycols gives polyesters; esters formed by reaction with monoalcohols are important plasticizers and lubricants (Bolton, 1995).

3.4. Glutaric acid

Glutaric acid occurs in washings from fleece and, together with malonic acid, in the juice of unripened sugar beet.

Production: Glutaric acid is obtained from cyclopentane by oxidation with oxygen and cobalt (III) catalysts or by ozonolysis; and from cyclopentanol – cyclopentanone by oxidation with oxygen and $Co(CH_3CO_2)_2$, with potassium peroxide in benzene, or with N_2O_4 or nitric acid. Like succinic acid, glutaric acid is formed as a byproduct during oxidation of cyclohexanol – cyclohexanone. Other production methods include reaction of malonic ester with acrylic acid ester, carbonylation of γ-butyrolactone, oxidation of 1,5-pentanediol with N_2O_4, and oxidative cleavage of γ-caprolactone.

Uses: The applications of glutaric acid, e.g., as an intermediate, are limited. Its use as a starting material in the manufacture of maleic acid has no commercial importance.

3.5. Adipic acid

Adipic acid, hexanedioic acid, 1,4-butanedicarboxylic acid, $C_6H_{10}O_4$, M_r 146.14, $HOOCCH_2CH_2CH_2CH_2COOH$ [124-04-9], is the most commercially important aliphatic dicarboxylic acid. It appears only sparingly in nature but is manufactured worldwide on a large scale. The historical development of adipic acid was reviewed in 1997 (Luedeke, 1997)

Physical properties: Adipic acid is isolated as colorless, odorless crystals having an acidic taste. It is very soluble in methanol and ethanol, soluble in water and acetone, and very slightly soluble in cyclohexane and benzene. Adipic acid crystallizes as monoclinic prisms from water, ethyl acetate, or acetone/petroleum ether.

Chemical properties: Adipic acid is stable in air under most conditions, but heating of the molten acid above 230 – 250 °C results in some decarboxylation to give cyclopentanone [120-92-3], *bp* 131 °C. The reaction is markedly catalyzed by salts of metals, including iron, calcium, and barium. The tendency of adipic acid to form a cyclic anhydride by loss of water is much less pronounced compared to glutaric or succinic acids.

Adipic acid readily reacts at one or both carboxylic acid groups to form salts, esters, amides, nitriles, etc. The acid is quite stable to most oxidizing agents, as evidenced by its production in nitric acid. However, nitric acid will attack adipic acid autocatalytically above 180 °C, producing carbon dioxide, water, and nitrogen oxides.

Use: Adipic acid has been used in the manufacture of mono- and diesters as well as polyamides. Nylon 6,8 is obtained by reaction of suberic acid with hexamethylenediamine, and nylon 8,8 by reaction with octamethylenediamine. Polyamides of adipic acid with diamines such as 1,3-bis(aminomethyl)benzene, 1,4(bisaminomethyl)cyclohexane, and bis(4-aminocyclohexyl)methane are also of commercial interest. Esters of adipic acid with mono- and bifunctional alcohols are used as lubricants.

4. Dicarboxlic acids distributions in the atmosphere

Numerous organic compounds significantly contribute to the aerosol load of the atmosphere and thus to the radiative forcing of climate. Among others the influence of organic aerosol on cloud droplet formation is a key point in evaluating effects of anthropogenic emissions on climate. In contrast to sulfate more uncertainties exist about organics and in particular for secondary organic aerosol species which are more oxygenated and hygroscopic than primary organic species (Saxena and Hildemann, 1996). Among oxygenated organic species, dicarboxylic acids are probably the best quantified species, though they represent a small fraction of the total organic mass (Kawamura and Ikushima, 1993). Glutaric and malonic acid the atmosphere have potential to increase the cloud condensation nuclei (CCN) activation of major inorganic aerosol such as ammonium sulfate (Cruz and Pandis, 1998). These findings suggest a potentially important role played by dicarboxylic acids on radiative forcing and stimulate their studies since the sources of diacids in the atmosphere remain poorly understood and quantified.

Whatever the region; urban and continental, or remote marine (see Figure 1 which carried out from Table 3), oxalic acid (C_2: HOOCCOOH) is always found to be the most abundant diacid followed by succinic (C_4: HOOC(CH$_2$)$_2$COOH) and/or malonic (C_3: HOOCCH$_2$COOH) acid with concentrations of several hundreds of nanograms per cubic meter in urban and continental regions (Kawamura and Ikushima, 1993; Kawamura and Kaplan, 1987) to a few tens of nanograms per cubic meter in remote marine boundary layer (Kawamura and Sakagushi, 1999; Sempere and Kawamura, 2003). In Europe, the most continuous study of diacids was conducted over one year by Limbeck et al., (2005) at Vienna, Austria. Although available data on diacids are more sparse at midlatitudes in Europe, they tend to show that oxalic acid levels at nonurban or rural sites are not considerably different from those at urban sites (Limbeck and Puxbaum, 1999; Rohrl and Lammel, 2001).

Motor exhausts have been proposed to be primary sources of oxalic, malonic, succinic, and glutaric (C_5: HOOC(CH$_2$)$_3$COOH) acids (Grosjean et al., 1978; Kawamura and Kaplan, 1987). Some of these diacids are also emitted by wood burning, particularly malonic acid (pine wood) and succinic acid (oak wood) (Rogge et al., 1991; Rogge et al., 1993). Note that until now no direct source of malic (hydroxysuccinic: hC_4: HOOCCH$_2$CHOHCOOH) and tartaric (dihydroxysuccinic: dhC_4: HOOC(CHOH)$_2$COOH) acids has been identified.

Glutaric, succinic, and adipic (C_6: HOOC(CH$_2$)$_4$COOH) acids have been identified in laboratory studies (Hatakeyama et al., 1985) as secondary organic aerosol products of the reaction of O$_3$ with cyclohexene, a symmetrical alkene molecule similar to monoterpenes emitted by the biosphere. Hatakeyama et al. (1985) also suggested that malonic and oxalic acids are also produced in the cyclohexene-ozone system.

Unsaturated fatty acids with a double bond at the C₉ position like cis-9-octadecenoic (oleic) acid are oxidized into C₉ diacid (azelaic acid) and other products hereafter mainly oxidized into shorter diacids hahah(Kawamura and Ikushima, 1994; Kawamura and Kaplan, 1987; Kawamura et al., 1985). These unsaturated acids which are abundant in marine phytoplankton and terrestrial higher plant leaves are also emitted by anthropogenic sources such as meat cooking (Rogge, 1991; Rogge et al., 1998) and wood burning processes (Rogge et al., 1998).

Warneck suggested that in the marine atmosphere clouds generate oxalic acid from glyoxal formed by oxidation of acetylene and glycolaldehyde formed by oxidation of ethane (Warneck, 2000). Note that along these processes glyoxylic acid (CHOCOOH) represents a key intermediate (see figure 3) whereas diacids other than oxalic acid are not produced. The formation of dicarboxylic acids in the continental atmosphere (Ervens et al., 2004a) involves production of glyoxal from toluene and of glycolaldehyde from isoprene as well as aqueous phase reactions of adipic and glutaric acids produced by oxidation of cyclohexene. Recently more literature has become available on the formation of oxalic acid that includes also the oxidation of methylglyoxal, an oxidation product of toluene and isoprene, via intermediate steps involving pyruvic and acetic acids (Lim et al., 2005). Since this diacid production pathway also forms oligomers, the knowledge of the sources of diacids is also of importance for the understanding of secondary organic aerosol formation.

The relative contribution of primary and secondary sources of diacids in the atmosphere remains poorly understood. Even though it is agreed that they are likely to be mainly secondary in origin it is not known in which proportion their precursors come from anthropogenic and biogenic sources.

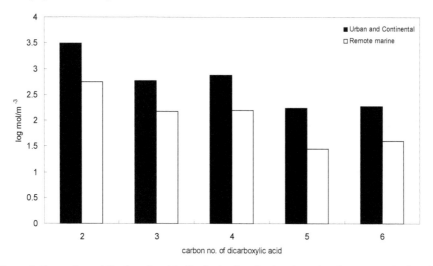

Figure 2. Comparison of dicarboxylic acids distribution in urban/continental and remote marine based on the data collection on table 3

References	Location	Oxalic	Malonic	Succinic	Glutaric	Adipic
(Grosjean et al., 1978)	New York	0	3.6	21.8	17.2	13.2
(Grosjean et al., 1978)	New York	0	3.9	24.9	23.2	11.6
(Kawamura and Kaplan, 1987)	West LA	6.38	1.58	1.96	0.6	2.22
(Kawamura and Kaplan, 1987)	West LA	2.12	0.4	0.66	0.22	0.94
(Kawamura and Kaplan, 1987)	West LA	8.13	0.72	2.34	0.66	3.31
(Kawamura and Kaplan, 1987)	West LA	8.65	1.45	2.37	0.74	0.49
(Kawamura and Kaplan, 1987)	Down Town LA	6.21	0.71	1.19	0.52	0.1
(Kawamura and Kaplan, 1987)	Down Town LA	6.6	0.76	1.84	0.52	0.2
(Kawamura and Kaplan, 1987)	Down Town LA	8.31	1.22	2.13	0.83	0.63
(Sempere and Kawamura, 1994)	Tokyo	29.65	6.69	13.18	3.72	6.66
(Sempere and Kawamura, 1994)	Tokyo	58.89	20.29	28.82	7.54	6.79
(Sempere and Kawamura, 1994)	Tokyo	330	141.3	161.1	4.15	2.91
(Limbeck and Puxbaum, 1999)	South Africa	193	142	58	8.8	7.9
(Limbeck and Puxbaum, 1999)	Sonblick Observatory	153	22	14	2.7	4.4
(Limbeck and Puxbaum, 1999)	Vienna	340	244	117	26	117
(Kawamura and Watanabe, 2004)	Tokyo	357	71.4	73.4	23.1	25.8
(Kawamura and Watanabe, 2004)	Tokyo	157	44	41	11	13
(Kawamura and Watanabe, 2004)	Tokyo	186	40.5	47.4	18.2	14.2
(Rohrl and Lammel, 2001)	Helsinki	0	0	30	0	0
(Ho et al., 2006)	Hong Kong (Road)	478	89.1	71.88	20	10.7
(Ho et al., 2006)	Hong Kong (Road)	268	47.6	33	6.95	12.7
(Hsieh et al., 2007)	Tainan,Taiwan	574	65.8	101	43	13.2
(Hsieh et al., 2007)	Tainan,Taiwan	432	34.2	87.9	10.3	8.8
(Limbeck et al., 2005)	Vienna, Austria	99.6	34	37	7.7	3.3
(Limbeck et al., 2005)	Vienna, Austria	66.2	38.6	30.8	6.6	3.2
(Limbeck et al., 2005)	Vienna, Austria	63.1	21.5	31.2	5.6	2.5
(Limbeck et al., 2005)	Mt Rax, Austria	34.5	9.1	16.4	2.3	0.8
(Limbeck et al., 2005)	Mt Rax, Austria	26.4	6.9	14.9	2.3	4.3
(Limbeck et al., 2005)	Mt Rax, Austria	32.6	16.4	22.4	3	1.7
(Decesari et al., 2006)	Rondonia, Brazil	194.7	73.1	123.5	23.5	14.5
(Decesari et al., 2006)	Rondonia, Brazil	793.3	56.8	210.2	32.1	12.6
(Decesari et al., 2006)	Rondonia, Brazil	937.9	128.5	423.9	34.7	21.2
(Decesari et al., 2006)	Rondonia, Brazil	1260	476.5	667.2	121.1	97.4
(Wang et al., 2006)	Hong Kong (Tunnel)	505	69.4	85.2	20.9	26.4
(Wang et al., 2006)	Hong Kong (Tunnel)	221	34.5	32.7	14.7	13.5
(Wang et al., 2006)	Hong Kong (Tunnel)	234	42	51.4	17.1	24.7
(Wang et al., 2006)	Hong Kong (Tunnel)	312	59.7	62.9	16.7	15.5
(Wang et al., 2006)	Hong Kong (Tunnel)	633	59.3	95.1	30.3	25.9

(a)

References	Location	Oxalic	Malonic	Succinic	Glutaric	Adipic
(Kawamura and Kaplan, 1987)	Green House LA	1.31	0.3	0.29	0.04	0.1
(Kawamura and Kaplan, 1987)	Green House LA	2.83	0.14	0.86	0	0.22
(Kawamura and Sakagushi, 1999)	North Pacific	44.7	23.2	19.5	2.57	3.08
(Kawamura and Sakagushi, 1999)	North Pacific	8.73	2.18	2.16	0.61	1.26
(Kawamura and Sakagushi, 1999)	North Pacific	10.6	1.98	2.22	0.23	2.12
(Kawamura and Sakagushi, 1999)	North Pacific	28.6	12.8	13	1.84	1.34
(Kawamura and Sakagushi, 1999)	North Pacific	667	189	93	20.1	4.9
(Kawamura and Sakagushi, 1999)	North Pacific	190	38.6	16.7	10.2	2.76
(Kawamura and Sakagushi, 1999)	North Pacific	88.5	34.5	21.6	4.72	6.04
(Kawamura and Sakagushi, 1999)	North Pacific	24.9	5.66	10.1	1.87	1.67
(Kawamura and Sakagushi, 1999)	North Pacific	10	2.12	1.52	0.32	0.43
(Kawamura and Sakagushi, 1999)	North Pacific	18.3	3.45	4.02	0.62	0.46
(Kawamura and Sakagushi, 1999)	North Pacific	25.5	5.93	2.99	0.65	0.4
Kawamura (1996)	Antarctic	1.59	0.13	0.63	0.31	0.49
Kawamura (1996)	Antarctic	3.12	0.38	5.77	0.58	0.85
Kawamura (1996)	Antarctic	3.26	0.52	1.18	0.34	0.33
Kawamura (1996)	Antarctic	10.29	2.69	61.53	2.26	1.81
Narukawa(1999)	Indonesia	2200	800.3	1090	310	350
Narukawa(1999)	Indonesia	225	18.4	123	30	40
Khwaja (1994)	semi urban site NY	308	84	55	12	89
Khwaja (1994)	semi urban site NY	245	92	106	16.3	101
Khwaja (1994)	semi urban site NY	118	165	107	15	40
Khwaja (1994)	semi urban site NY	58	81	129	20	21
Khwaja (1994)	semi urban site NY	298	96	90	23	31
Khwaja (1994)	semi urban site NY	1	43	0.5	39	20
Khwaja (1994)	semi urban site NY	360	88	167	46	50
Sempere (2003)	Western Pacific	428.5	78.6	33.4	7.6	7.2
Rohrl (2002)	rural(I)	0	0	14	0	0
Rohrl (2002)	rural(II)	0	0	8.8	0	0
Rohrl (2002)	rural(III)	0	0	18	0	0
(Kawamura et al., 2007)	Canadian arctic	9.89	2.74	2.16	0.54	0.51
(Kawamura et al., 2007)	Canadian arctic	8.3	2.87	1.44	0.37	0.26
(Kawamura et al., 2007)	Canadian arctic	5.26	1.67	1.08	0.22	0.27
(Narukawa et al., 2002)	Arctic,Alert	23.5	5.03	3.21	1.21	0.54
(Narukawa et al., 2002)	Arctic,Alert	40.09	11.6	15.67	2.16	0.55
(Mochida et al., 2007)	North Pacific,ACE	600	110	52	8.9	2
(b)						

Table 3. Summary of aerosol dicarboxylate concentration (ng m^{-3}) in urban/continental (a) remote marine (b) locations

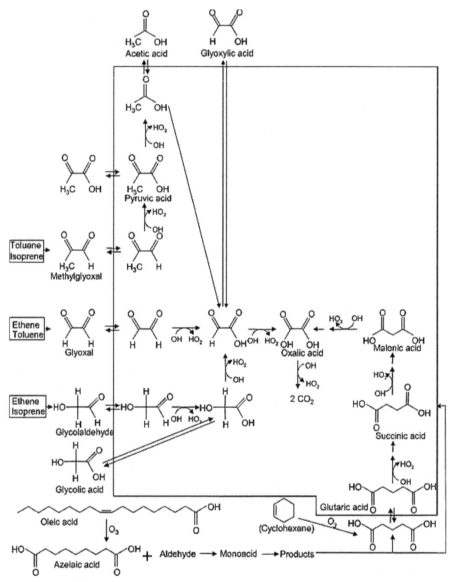

Figure 3. Multiphase organic chemistry producing C_2–C_5 diacids from key biogenic and anthropogenic precursors. The box refers to the aqueous phase. The figure is mainly adapted from (Ervens et al., 2004b) with modifications to account for the reaction pathway methylglyoxal/pyruvic acid/acetic acid/glyoxylic acid suggested by (Lim et al., 2005). In addition to cyclohexene used by (Ervens et al., 2004b)as a model compound for symmetrical alkenes, following (Legrand et al., 2007) we also report the oleic acid degradation into azelaic, C_4 and C_5 diacids.

Author details

Mohd Zul Helmi Rozaini

School of Environmental Sciences, University of East Anglia, Norwich, Norfolk, UK
Department of Chemical Sciences, University Malaysia Terengganu, Kuala Terengganu,
Terengganu, Malaysia

5. References

Adam, P.J., Seinfeld, J.H. and Koch, D.M., 1999. Global concentration of tropospheric sulfate, nitrate and ammonium aerosol simulated in a general circulation model. J. Geophys. Res, 104: 13791.

Bolton, G.L., 1995. Encyclopedia of Reagents for Organic Synthesis, 5. John Wiley & Sons, New York, 3213 – 3222. pp.

Brimblecombe, P., 1987. The big smoke: A history of air pollution in London since medieval times. Methuen, New York.

Cachier, H., Bremond, M.P., and and Buatmenard, P., 1989. Carboneceous Aerosols from Different Tropical Biomass Burning Sources. Nature, 340: 371-373.

Charlson, R.J. and Heintzenberg, 1995. Aerosols as acaouse of uncertainty in climate forecast: Report of Dahlem Workshop on Aerosl Forcing of Climate. Wiley & Sons, Chichester, pp. 1-10.

Clarke, H.T.a.D., A. W.,, 1986. Organic Synthesis. J. Wiley and sons, New York, pp. 421-425.

Cornell, S.E., Jickells, T.D., and and Cape, J.N., 2003. Organic nitrogen deposition on land and coastal environments; a review of methods and data. Atmospheric Envionment, 37: 2173-2191.

Cruz, C.N. and Pandis, S.N., 1998. The effect of organic coatings on the cloud condensation nuclei activation of inorganic atmosphere aerosol. Journal of Geophys. Res., 103(D11): 13111-13123.

Cruz, C.N., Pandis, S.N.,, 1998. The effect of organic coatings on the cloud condensation nuclei activation of inorganic atmosphere aerosol. Journal of Geophys. Res., 103(D11): 13111-13123.

Decesari, S., Facchini, M.C., Fuzzi, S., & and Tagliavini, E., 2000. Characterization of water-soluble compounds in atmospheric aerosol: A new approach. Journal of Geophysiscal research, 105: 1481-1489.

Decesari, S. et al., 2002. Water soluble organic compounds formed by oxidation of soot. Atmospheric Environment, 36(11): 1827-1832.

DEFRA, 2005. UK Department for Environment Food & Rural Afffairs, The Quality Strategy for England, Scotland, Wales and Northern Island (Rep. 01EP0538), London,UK.

Dollimore, D., 1987. Thermochimica Acta, 117: 331-363.

Dutot, A.L., Rude, J., and and Aumont, B., 2003. Neutral network method to estimate teh aqueous rate constant for the OH reactions with organic compounds. Atmospheric Envionment, 37: 269-276.

Ervens, B., Feingold, G., Clegg, S.L. and Kreidenweis, S.M., 2004a. A modeling study of aqueous production of dicarboxylic acids: 2. Implications for cloud microphysics. Journal of Geophysical Research D: Atmospheres, 109(15).

Ervens, B., Feingold, G., Frost, G.J. and Kreidenweis, S.M., 2004b. A modeling of study of aqueous production of dicarboxylic acids: 1. Chemical pathways and speciated organic mass production. Journal of Geophysical Research D: Atmospheres, 109(15).

Fenger, J., 1999. Urban air quality. Atmospheric Environment, 33(29): 4877-4900.

Fisseha, R., Dommen, J. and Sax, M., 2004. idenfification of organic aerosol and the acids in secondary corresponding gas phase from chamber experiments. Analytical Chemistry, 76: 6335-6540.

Grosjean, D., Van Cauwenberghe, K. and Schmid, J.P., 1978. Identification of C3-C10 aliphatic dicarboxylic acids in airborne particulate matter. Environmental Science and Technology, 12(3): 313-317.

Grossi, C.M., and P.Brimblecombe, 2002. The effect of atmospheric pollution on building materials. Journal De Physique Iv, 12(PR10): 197-210.

Hatakeyama, S., T., et al., 1985. Ozone-cyclohexene reaction in air: Quantitative analyses of particulate products and the reaction mechanism. Environ. Sci. Technol, 19,: 935–942.

Jacob, D.J., 1999. Introduction to a atmospheric chemistry. Princeton University, New Jersy.

Kawamura, K., and and Sakagushi, F., 1999. Molecular distribution of water soluble dicarboxylic acids in marine aerosols over the pacific ocean including tropics. J. Geophys. Res, 104: 3501-3509.

Kawamura, K. and Ikushima, K., 1993. Seasonal changes in the distribution of dicarboxylic acids in the urban atmosphere. Environmental Science & Technology, 27(10): 2227-2235.

Kawamura, K. and Ikushima, K., 1994. Seasonal changes in the distribution of dicarboxylic acids in the urban atmosphere. Environ. Sci. Technol, 27: 2227-2235.

Kawamura, K. and Kaplan, I.R., 1987. Dicarboxylic acids generated by thermal alteration of kerogen and humic acids. Geochimica et Cosmochimica Acta, 51: 3201-3207.

Kawamura, K., Kasukabe, H. and Barrie, L.A., 1996. Source and reaction pathways of dicarboxylic acids, ketoacids and dicarbonyls in arctic aerosols: One year of observations. Atmospheric Environment

Kawamura, K., Ng, L.L. and Kaplan, I.R., 1985. Determination of organic acids (C1-C10) in the atmosphere, motor exhausts, and engine oils. Environmental Science and Technology, 19(11): 1082-1086.

Kerminen, V.-M. et al., 2000. Low molecular weight dicarboxylic acids in an urban and rural atmosphere. Journal of Aerosol Science, 31(3): 349-362.

Khwaja, H.A., 1995. Atmospheric concentrations of carboxylic acids and related compounds at a semiurban site. Atmospheric Environment, 29(1): 127-139.

Legrand, M. et al., 2007. Origin of C2–C5 dicarboxylic acids in the European atmosphere inferred from year-round aerosol study conducted at a west-east transect. J. Geophys. Res.,, 112: D23S07, doi:10.1029/2006JD008019.

Lim, H.-J., A. G. Carlton, a. and Turpin, B.J., 2005. Isoprene forms secondary organic aerosol through cloud processing: Model simulations. Environ. Sci. Technol., 39: 4441–4446.

Limbeck, A., Kraxner, Y. and Puxbaum, H., 2005. Gas to particle distribution of low molecular weight dicarboxylic acids at two different sites in central Europe (Austria). Journal of Aerosol Science, 36(8): 991-1005.

Limbeck, A. and Puxbaum, H., 1999. Organic acids in continental background aerosols. Atmospheric Environment, 33(12): 1847-1852.

Luedeke, V., 1997. Encyclopedia of Chemical Processing and Design. In: W.C. J. McKetta (Editor). Marcel Dekker Inc, New York, pp. 128 – 146.

Moreno, T., Gibbons, W., Jones, T. and Richards, R., 2003. The geology of ambient aerosols: Characterising urban and rural/coastal silicate PM10-2.5 and PM2.5 using high-volume cascade collection and scanning electron microscopy. Atmospheric Environment, 37(30): 4265-4276.

MOSTI, M.O.S.a.I.M., 2000. Environmental Quality (Standard for Particulate Matter) Regulations, Kuala Lumpur.

Narukawa, M., Kawamura, K., Takeuchi, N. and Nakajima, T., 1999. Distribution of dicarboxylic acids and carbon isotopic compositions in aerosols from 1997 Indonesian forest fires. Geophysical Research Letters, 26(20): 3101-3104.

Novokov, T. and Penner, J.E., 1993. Large Contribution of Organic Aerosols to Cloud-Condensation-Nuclei Concentration. Nature, 365: 823-826.

Preining, O., 1993. Global Climate Change Due to Aerosols. In: H.C. N. (Editor), Global Atmospheric Chemical Change. Elsevier Applied science, London and New York, pp. 93-122.

Puxbaum, H. and Limbeck, A., 2000. Scavenging of Organic aerosol Constituents in Supercooled clouds, 13th International Conference on Cloud Precipitation, Reno,Nevada.

Rogge, W.F., Lynn M. Hildemann, Monica A. Mazurek, Glen R. Cass, a. and Simoneit, B.R.T., 1991. Sources of fine organic aerosol. 1. Charbroilers and meat cooking operations. Environ. Sci. Technol, 25: 1112-1125.

Rogge, W.F., Lynn M. Hildemann, Monica A. Mazurek, Glen R. Cass, and Bernd R. T. Simoneit, 1991. Sources of fine organic aerosol. 1. Charbroilers and meat cooking operations. Environ. Sci. Technol, 25: 1112-1125.

Rogge, W.F., Mazurek, M.A., Hildemann, L.M., Cass, G.R. and Simoneit, B.R.T., 1993. Quantification of urban organic aerosols at a molecular level: Identification, abundance and seasonal variation. Atmospheric Environment - Part A General Topics, 27 A(8): 1309-1330.

Rogge, W.F., Mazurek, M.A., Hildemann, L.M., Cass, G.R. and Simoneit, B.R.T., 1998. Quantification of urban organic aerosols at a molecular level: Identification, abundance and seasonal variation. Atmospheric Environment - Part A General Topics, 27 A(8): 1309-1330.

Rohrl, A. and Lammel, G., 2001. Low-Molecular Weight Dicarboxylic Acids and Glyoxylic Acid: Seasonal and Air Mass Characteristics. Environ. Sci. Technol., 35(1): 95-101.

Rudloff, J. and Cölfen, H., 2004. Langmuir, 20: 991.

Saxena, P., and and Hildemann, L.M., 1996. Water-Soluble Organics in Atmospheric
    Particles: A critical Review of the Literature and Application of Thermodynamics to
    Identify Candidate Compounds. J. Atmos. Chem, 24: 57-109.
Saxena, P. and Hildemann, L.M., 1997. Water Absorption by Organics: Survey of Laboratory
    Evidence and Evaluation of UNIFAC for Estimating Water Activity. Environ. Sci.
    Technol., 31(11): 3318-3324.
Sempere, R. and Kawamura, K., 2003. Trans-hemispheric contribution of C2-C10 a,w-
    dicarboxylic acids, and related polar compounds to water-soluble organic carbon in the
    western Pacific aerosols in relation to photochemical oxidation reactions. Global
    Biogeochemical Cycles, 17(2): 38-1.
Simoneit, B.R.T. and Mazurek, D.A., 1982. Organic matter of the troposphere II: Natural
    background of biogenic lipid matter in aerosols over the rural western United States.
    Atmos Environment, 16: 2139-2159.
Tedetti, M., Kawamura, K., Charriere, B., Chevalier, N. and Sempere, R., 2006.
    Determination of Low Molecular Weight Dicarboxylic and Ketocarboxylic Acids in
    Seawater Samples. Anal. Chem., 78(17): 6012-6018.
Warneck, P., 2000. Chemistry of the Natural Atmosphere. Second Edition, Oxford, AP
    publication.
Yu, L., Shulman, M., Kopperud, R. and Hildemann, L., 2005. Characterization of Organic
    Compounds Collected during Southeastern Aerosol and Visibility Study: Water-Soluble
    Organic Species. Environ. Sci. Technol., 39(3): 707-715.
Yu, S., 2000. Role of organic acids (formic, acetic, pyruvic and oxalic) in the formation of
    cloud condensation nuclei (CCN): a review. Atmospheric Research, 53(4): 185-217.
Yusunov, D., Tukhtaev, S., and and Semenova, L.N., 1972. Deposited Doc.,, VINITI: 4612-72.
Zappoli, S. et al., 1999. Inorganic, organic and macromolecular components of fine aerosol in
    different areas of Europe in relation to their water solubility. Atmospheric
    Environment, 33: 2733–2743.

Chapter 3

Effects of Inorganic Seeds on Secondary Organic Aerosol (SOA) Formation

Biwu Chu, Jingkun Jiang, Zifeng Lu, Kun Wang, Junhua Li and Jiming Hao

Additional information is available at the end of the chapter

1. Introduction

Atmospheric aerosol has significant influences on human health (Kaiser, 2005), visibility degradation (Cheng et al., 2011), and climate change (Satheesh and Moorthy, 2005). It was found that organic aerosols (OA) was the most abundant component of atmospheric aerosol (He et al., 2001) and more than 50% of the total OA are secondary organic aerosols (SOA) (Duan et al., 2005). SOA are produced from the oxidation of volatile organic compounds (VOCs) followed by gas-particle partitioning of the semivolatile organic products. Among the various VOCs, aromatic hydrocarbons are one type of SOA precursors which have drawn the most attention due to their abundance in the air and high SOA contribution to urban atmospheres (Lewandowski et al., 2008). Toluene and *m*-xylene are the two of the most abundant aromatic hydrocarbon species.

The detailed mechanism and controlling factors of SOA formation are not fully understood yet, which leads to the lower SOA level prediction from air quality models than the ambient measurements (Volkamer et al., 2006). Using smog chamber, SOA formation process can be investigated under controlled experimental conditions. Series of smog experiments have been conducted by different research groups to investigate the effects of background seed aerosols on SOA formation (Cao and Jang, 2007, Czoschke et al., 2003, Gao et al., 2004, Jang et al., 2002, Liggio and Li, 2008). Increased SOA formation and SOA yields were observed with the presence of acid seed aerosols. The effects of acidic seeds suggest that aerosol phase reactions may play an important role on SOA formation (Jang et al., 2002). Interactions between the organic and inorganic components of aerosols are important for further understanding the SOA formation process. Most research concludes that acid-catalyzed aerosol-phase reactions generate additional aerosol mass due to the production of oligomeric products with large molecular weight and extremely low volatility (Cao and Jang, 2007, Czoschke et al., 2003, Gao et al., 2004) and, therefore, enhance SOA formation.

Uptake of semivolatile organic products to acidic sulfate aerosols was also found contributing to enhance SOA formation (Liggio and Li, 2008). In these studies, $(NH_4)_2SO_4$ or H_2SO_4 seed aerosols were widely used to study the effect of particle acidity on SOA formation from both biogenic and aromatic hydrocarbons.

Atmospheric aerosols always have a very complex composition. Studying the effects of $(NH_4)_2SO_4$ or H_2SO_4 seed aerosols did not draw the whole picture of the role that inorganic seed aerosols play in SOA formation. Metal-containing aerosols are important components of the atmosphere. Calcium and iron are the most abundant metal species in atmospheric aerosols and the average concentration of them in Beijing could be as high as about 1.2 μg m^{-3} and 1.1 μg/m^3 in $PM_{2.5}$ (He et al., 2001) respectively. In this study, we tested the effect of different inorganic seeds on SOA formation using a smog chamber. Two aromatic hydrocarbon precursors toluene and m-xylene are used. Effects of various inorganic seeds, including neutral inorganic seed $CaSO_4$, acidic seed $(NH_4)_2SO_4$, transition metal contained inorganic seeds $FeSO_4$ and $Fe_2(SO_4)_3$, and a mixture of $(NH_4)_2SO_4$ and $FeSO_4$, were examined during m-xylene or toluene photooxidation with the presence of nitrogen oxides (NO_x).

2. Experimental section

The experiments were carried out in a smog chamber which was described in detail in Wu et al. (Wu et al., 2007). The 2 m^3 cuboid reactor, with a surface-to-volume ratio of 5 m^{-1}, was constructed with 50 μm-thick FEP-Teflon film (Toray Industries, Inc. Japan). The reactor was located in a temperature controlled room (Escpec SEWT-Z-120), with a constant temperature between 10 and 60 °C (± 0.5 °C). The reactor was irradiated by 40 black lights (GE F40T12/BLB, peak intensity at 365 nm). Based on the equilibrium concentrations of NO, NO_2 and O_3 in a photo-irradiation experiment of an NO_2/air mixture, the NO_2 photolysis rate was calculated at approximately 0.21 min^{-1}, using a method described by Takekawa et al. (2000, 2003).

Prior to each experiment, the chamber was flushed for about 40 h with purified air at a flow rate of 15 L/min. In the first 20 hours, the chamber was exposed to UV light at 34 °C. In the last several hours of the flush, humid air was introduced to obtain the target relative humidity (RH).

Seed aerosols were generated by atomizing salt solutions using a constant output atomizer (TSI Model 3076). To avoid hydrolysis and precipitation in the $Fe_2(SO_4)_3$ salt solution, as little sulfuric acid as possible was added to the solution. What's more, for generating internally mixed seed aerosols, a mixed solution of $FeSO_4$ and $(NH_4)_2SO_4$, in which the concentration ratio of $FeSO_4$ to $(NH_4)_2SO_4$ is 1:5, was used. The generated aerosols were passed through a diffusion dryer (TSI Model 3062) to remove water and a neutralizer (TSI Model 3077) to bring the aerosols to an equilibrium charge distribution. The hydrocarbon, NO and NO_2 were carried by purified dry air into the chamber. The concentrations were continuously monitored at a measurement point in the reactor until they were stable, ensuring the components in the reactor were well mixed. The experiment was then conducted for 6 hours with the black lights on.

A gas chromatograph (GC, Beifen SP-3420) equipped with a DB-5 column (30 m×0.53 mm×1.5 mm, Dikma) and flame ionization detector (FID) measured the concentration of the hydrocarbon every 15 min. NO_x and O_3 were monitored with an interval of 1 min by a NO_x analyzer (Thermo Environmental Instruments, Model 42C) and an O_3 analyzer (Thermo Environmental Instruments, Model 49C), respectively. Size distribution of particle matter (PM) was measured by a scanning mobility particle sizer (SMPS, TSI 3936) in the range of 17-1000 nm with a 6-min cycle. The volume concentration of aerosols was estimated from the measured size distribution by assuming the particles were geometrically spherical and nonporous.

3. Results and discussion

3.1. Estimating the generated SOA mass (M_o)

Due to deposition of particles on the Teflon film, the measured aerosol concentration had to be corrected. Takekawa et al. (2003) developed a particle size-dependent correction method, in which the aerosol deposition rate constant ($k(d_P)$, h^{-1}) is a four-parameter function of particle diameter (d_P, nm), as shown in equation (1):

$$k(d_p) = a \bullet d_p^{b} + c \bullet d_p^{d} \tag{1}$$

The resulting $k(d_P)$ values for different d_P (40-700 nm) were determined by monitoring the particle number decay under dark conditions at low initial concentrations (<1000 particles cm^{-3}) to avoid serious coagulation. Based on more than 500 sets of $k(d_P)$ values (d_P ranges from 40 to 700 nm), the optimized values of parameter a, b, c, and d were calculated to be $6.46×10^{-7}$, 1.78, 13.2, and -0.957, respectively. It should be noted that the estimation of deposited aerosol concentrations using this method might introduce some error (Takekawa et al., 2003) because some scatter was recognized when fitting $k(d_P)$ values into equation (1). To reduce error due to wall deposition, SOA yields were calculated when the measured particle concentration reached its maximum in the experiments because deposited aerosols were a greater proportion of the aerosol concentration change in the reactor after that time.

Several researchers have measured SOA density, providing an estimated range of 0.6-1.5 g cm^{-3} (Bahreini et al., 2005, Poulain et al., 2010, Qi et al., 2010, Song et al., 2007, Yu et al., 2008). In our study, we used a unit density (1.0 g cm^{-3}) to calculate SOA mass concentrations. This follows the approach used in Takekawa et al. (2003) and Verheggen et al. (2007).

3.2. Calculation of SOA yields

The fractional SOA yield (Y), defined as the ratio of the generated organic aerosol concentration (M_o) to the reacted hydrocarbon concentration (ΔHC), was used to represent the aerosol formation potential of the hydrocarbon (Pandis et al., 1992). Odum et al. (1996) developed a gas/particle absorptive partitioning model to describe the phenomenon that Y largely depends on the amount of organic aerosol mass present. Equation (2) illustrates the relationship between SOA yield and organic aerosol mass concentration:

$$Y = \frac{\Delta M_o}{\Delta HC} = \frac{\sum_i A_i}{\Delta HC} = M_o \sum_i \frac{\alpha_i K_{om,i}}{1 + K_{om,i} M_o} \qquad (2)$$

In equation (2), i presents the serial number of the hydrocarbon reaction products, A_i, α_i and $K_{om,i}$ (m^3 μg^{-1}) are the aerosol mass concentration, the stoichiometric coefficient based on mass and the normalized partitioning constant for product i respectively. If we assume that all semi-volatile products can be classified into one or two groups, equation (2) can be simplified to a one-product model (i.e., i=1) or two-product model (i.e., i=2). Parameters (α and K_{om}) can be obtained by fitting the experimental SOA yield data with a least square method. Since numerous compounds are actually produced by the reaction of a hydrocarbon, parameters obtained by the simplified model only represent the overall properties of all products (Odum et al., 1996). A one-product model was proved sufficiently accurate to describe the relationship between aerosol yield and mass (Henry et al., 2008, Takekawa et al., 2003, Verheggen et al., 2007). Therefore, we used a one-product model for our experimental SOA yield data to quantify of the effects of inorganic seed aerosols on SOA formation.

3.3. Effects of CaSO₄ and (NH₄)₂SO₄ seed aerosols on SOA formation

To investigate the effects of neutral and acid aerosols on SOA formation in m-xylene photooxidation, $CaSO_4$ and $(NH_4)_2SO_4$ were selected as surrogates. Experimental conditions were listed in Table 1. Six seed-free experiments (Xyl-N1~6), three CaSO₄-introduced experiments (Xyl-CS1~3) and nine (NH₄)₂SO₄-introduced experiments (Xyl-AS1~9) were carried out. Among these experiments, some experiments have identical initial conditions except for the seed aerosols (i.e. experiments Xyl-N5, Xyl-CS2, Xyl-AS2, Xyl-AS3, Xyl-AS9). Comparing the temporal variation of NO and O_3 during these experiments with similar initial conditions (Figure 1), the results indicate that $CaSO_4$ and $(NH_4)_2SO_4$ seed aerosols have no significant effect on gas-phase reactions. This result is consistent with the findings of Kroll et al. (2007) and Cao and Jang (2007) that $(NH_4)_2SO_4$ and $(NH_4)_2SO_4/H_2SO_4$ seed aerosols had a negligible effect on hydrocarbon oxidation.

Similarly, by comparing the temporal variation particle concentrations (Figure 2) during the experiments with identical initial conditions except for the seed aerosols, the effects of $CaSO_4$ and $(NH_4)_2SO_4$ seed aerosols on SOA formation were identified. In Figure 2, PMcorrected was calculated from the measured PM concentrations plus wall deposit loss, and PM₀ was the seed aerosol concentration. The results indicate that the presence of neutral aerosols $CaSO_4$ (16-73μg m^{-3}) in the m-xylene/NOₓ photooxidation system have no significant effect on SOA formation. Experiments with the presence of acid aerosols $(NH_4)_2SO_4$ have different particle profiles according to the concentrations of the introduced $(NH_4)_2SO_4$ seed aerosol. In Figure 2, experiment Xyl-AS2 has similar particle profile with the seed-free experiment Xyl-N5, indicating that $(NH_4)_2SO_4$ seed aerosols have little effect on SOA formation when the initial concentration is low. However, when with high concentration of $(NH_4)_2SO_4$ seed aerosol introduced, SOA formation was enhanced (i.e. experiments Xyl-AS3 and Xyl-AS9) comparing with the seed-free experiment Xyl-N5. Comparing experiments Xyl-AS3 and Xyl-AS9, higher concentration of $(NH_4)_2SO_4$ seed aerosol resulted in higher SOA concentration. Therefore, the effects of $(NH_4)_2SO_4$ seed aerosol on SOA formation depend on its concentration.

Experiment no.	HC_0 (ppm)	PM_0 ($\mu m^3\,cm^{-3}$)	$PM_{0,S}$ ($cm^2\,m^{-3}$)	NO_0 (ppb)	$NO_{x,0}$ (ppb)	$HC_0/$ $NO_{x,0}$	M_0 ($\mu g\,m^{-3}$)	ΔHC (ppm)	Y (%)
Xyl-N1	0.92	0	-	72	148	6.2	66	0.27	5.6
Xyl-N2	1.26	0	-	102	209	6.0	92	0.32	6.7
Xyl-N3	1.74	0	-	137	276	6.3	122	0.39	7.3
Xyl-N4	1.68	0	-	132	272	6.2	125	0.39	7.5
Xyl-N5	2.00	0	-	161	333	6.0	148	0.45	7.6
Xyl-N6	2.51	0	-	182	381	6.6	191	0.54	8.3
Xyl-CS1	1.17	16	-	86	174	6.7	67	0.29	5.4
Xyl-CS2	2.03	43	-	167	343	5.9	148	0.46	7.6
Xyl-CS3	2.90	73	-	232	471	6.2	201	0.59	8.0
Xyl-AS1	2.85	11	3.5	208	420	6.8	208	0.57	8.6
Xyl-AS2	2.06	23	6.7	166	337	6.1	150	0.45	7.7
Xyl-AS3	2.15	47	11.0	162	326	6.6	169	0.45	8.8
Xyl-AS4	0.92	43	13.2	70	137	6.7	93	0.27	8.1
Xyl-AS5	1.73	45	12.8	132	272	6.4	165	0.40	9.7
Xyl-AS6	2.41	55	13.0	178	365	6.6	232	0.53	10.3
Xyl-AS7	0.92	63	16.7	70	143	6.4	110	0.26	10.1
Xyl-AS8	1.56	69	17.1	132	269	5.8	173	0.35	11.5
Xyl-AS9	2.07	74	17.1	166	348	6.0	249	0.47	12.3

Table 1. Initial experiment conditions and results for experiments with/without $CaSO_4$ or $(NH_4)_2SO_4$: initial m–xylene concentration (HC_0), initial seed aerosol mass concentration (PM_0), initial seed aerosol surface concentration ($PM_{0,S}$), initial NO_x concentrations (NO_0 and $NO_{x,0}$-NO_0), ratio of $HC_0/NO_{x,0}$, generated SOA mass (M_0), reacted hydrocarbon (ΔHC) , and SOA yield (Y)

Further analysis found that the effects of $(NH_4)_2SO_4$ seed aerosol on SOA yield were positively correlated with the surface concentration of $(NH_4)_2SO_4$ seed aerosols. To draw the SOA yield curves shown in Figure 3, the experiments were classified into different groups (experiment Xyl-AS3 was not classified into any group since the surface concentration of $(NH_4)_2SO_4$ seed aerosols in this experiment was different from others) by the surface concentration of $(NH_4)_2SO_4$ seed aerosols. The regression lines for each group (there was no regression line for experiments XylCS1~2 and Xyl-AS1~3 since they had similar SOA yield with the seed-free experiments) were produced by fitting the data of generated SOA mass (M_0) and SOA yield (Y) into a one-product partition model. As indicated in Figure 3, experiments with higher surface concentration of $(NH_4)_2SO_4$ seed aerosols had higher yield curves. As proposed by most research, acid-catalyzed aerosol-phase reactions (Cao and

Jang, 2007, Czoschke et al., 2003, Gao et al., 2004) and uptake of semivolatile organic products to acidic sulfate aerosols enhance SOA formation (Liggio and Li, 2008). The observed SOA formation enhancement could be related to the acid catalytic effect of $(NH_4)_2SO_4$ seeds on particle-phase surface heterogeneous reactions and the surface uptake of semivolatile organic products.

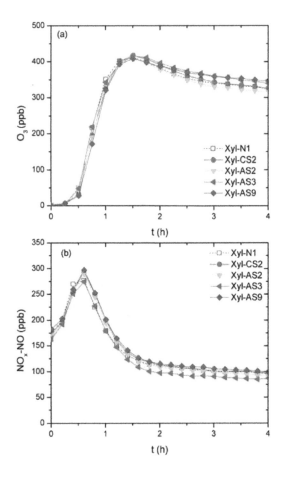

Figure 1. Temporal evolutions of O_3 (a) and NO_x-NO (b) concentration in experiments with/without $CaSO_4$ and $(NH_4)_2SO_4$ seed aerosols

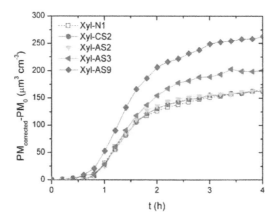

Figure 2. Temporal evolutions of generated particle concentration in experiments with/without CaSO₄ and (NH₄)₂SO₄ seed aerosols

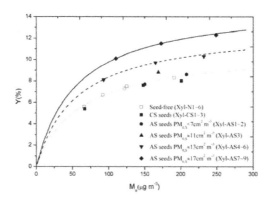

Figure 3. SOA yields (Y) from photooxidation of m-xylene versus organic aerosol mass (Mo) for experiments with/without CaSO₄ and (NH₄)₂SO₄ seed aerosols

3.4. Effects of $Fe_2(SO_4)_3$ and $FeSO_4$ seed aerosols on SOA formation

A seed-free experiment and three experiments with $Fe_2(SO_4)_3$ seed aerosols were carried out to investigate $Fe_2(SO_4)_3$ seed aerosols on phooxidation of toluene/NO_x. The four experiments had identical initial conditions except for the concentrations of the introduced $Fe_2(SO_4)_3$ seed aerosol. $Fe_2(SO_4)_3$ seed aerosols did not have obvious effects on SOA formation as shown in the temporal variation of $PM_{corrected}$–PM_0 concentrations in Figure 4. $Fe_2(SO_4)_3$ seed aerosols had no obvious effect on gas phase compounds in toluene/NO_x photooxidation either. A minimal amount of acid was added to the solution to generate $Fe_2(SO_4)_3$ seed aerosols. The introduced H^+ concentration was in the range of 0.0002-0.002 µg m⁻³ in the $Fe_2(SO_4)_3$-

introduced experiments. This is much lower than the H^+ concentration in the "non-acid" experiment by Cao and Jang (2007). Therefore, we presume the effect of the introduced sulfuric acid was negligible and $Fe_2(SO_4)_3$ seed aerosols did not have obvious effects on SOA formation in phooxidation of toluene/NO$_x$.

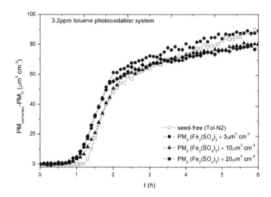

Figure 4. Variations of generated SOA mass as a function of time from toluene/NO$_x$ photooxidation with different concentrations of $Fe_2(SO_4)_3$ seed aerosols

We also conducted 18 irradiated toluene/NO$_x$ experiments with/without FeSO$_4$ seed aerosols. The conditions, generated SOA mass (M_o), and SOA yield (Y) are shown in Table 2. FeSO$_4$ seed aerosols had no obvious effect on gas phase compounds either, but significantly suppressed SOA formation. Figure 5 compares the temporal variation of particle concentrations during the 4.2 ppm toluene experiments (Exierments Tol-N3, Tol-FS1, Tol-FS3, Tol-FS8 and Tol-FS12) conducted under identical initial conditions except seed aerosol concentrations. Experiments with the presence of FeSO$_4$ seed aerosol generated less SOA than the seed-free experiment. And experiment with a higher FeSO$_4$ seed aerosol concentration generated less SOA than experiment with a lower FeSO$_4$ concentration. So the inhibited effect of FeSO$_4$ aerosols on SOA yield became stronger at higher concentrations of FeSO$_4$ seed aerosols. At other toluene/NO$_x$ photooxidation concentrations, we also found similar temporal variation of particle concentrations. However, as indicated in Table 2 and Figure 5, SOA yields of experiments Tol-FS1 and Tol-FS3 are similar to corresponding seed-free experiments of Tol-N3. These two seed-introduced experiments (as well as Tol-FS2) were conducted at the lowest ratio of FeSO$_4$ seed aerosol mass concentration to initial toluene mass concentration (FeSO$_4$/toluene) and did not show obvious effect on SOA formation comparing to their corresponding seed-free experiments. In these three experiments, the mass ratios of FeSO$_4$/toluene (assuming particle density to be 1.898 g cm^{-3}, density of FeSO$_4$·7H$_2$O, because of the lack of the information the amount of hydrate water) were calculated to be lower than 4.2×10^{-4}. It is possible that most of the ferrous iron was oxidized before significant SOA mass were generated since few FeSO$_4$ seed aerosols were introduced and high concentrations of oxidizing substances were generated during the toluene/NO_x photooxidation. Besides these three experiments with lowest $FeSO_4$/toluene mass ratio, $FeSO_4$ seed aerosols suppressed SOA formation relative to the corresponding seed-free experiments. And in our experiments, the suppress ratio could be as high as 60%, as calculated from Table 2.

Experiment No.	HC_0 ppm	PM_0 μm^3 cm^{-3}	NO_0 ppb	$NO_{x,0}-NO_0$ ppb	PM_0/HC_0	$HC_0/NO_{x,0}$ ppm ppm^{-1}	M_0 μg m^{-3}	ΔHC ppm	Y %
Tol-N1	1.10	0	50	51	0	11.0	26	0.20	3.8
Tol-FS4	1.08	1	51	50	5.1×10^{-4}	10.7	17	0.22	2.3
Tol-FS10	1.07	4	55	47	1.7×10^{-3}	10.6	14	0.20	2.2
Tol-FS14	1.09	10	48	49	4.4×10^{-3}	11.1	8	0.19	1.7
Tol-N2	3.30	0	165	160	0	10.2	90	0.48	5.0
Tol-FS5	3.21	4	160	162	6.1×10^{-4}	10.0	74	0.51	3.9
Tol-FS7	3.31	6	154	162	8.4×10^{-4}	10.5	72	0.56	3.5
Tol-FS9	3.19	11	164	157	1.5×10^{-3}	9.9	59	0.47	3.3
Tol-FS11	3.28	21	158	165	3.0×10^{-3}	10.2	36	0.51	1.9
Tol-N3	4.12	0	217	210	0	9.7	123	0.57	5.8
Tol-FS1	4.23	1	208	207	1.4×10^{-4}	10.2	105	0.57	5.0
Tol-FS3	4.25	4	208	213	4.2×10^{-4}	10.1	115	0.60	5.2
Tol-FS8	4.25	10	216	209	1.1×10^{-3}	10.0	81	0.55	4.0
Tol-FS12	4.23	27	213	210	3.0×10^{-3}	10.0	47	0.61	2.1
Tol-N4	6.10	0	287	293	0	10.5	189	0.96	6.3
Tol-FS2	6.05	5	295	306	3.5×10^{-4}	10.1	170	0.81	6.5
Tol-FS6	6.09	10	299	306	7.6×10^{-4}	10.1	140	0.88	4.8
Tol-FS13	6.03	41	296	310	3.2×10^{-3}	10.0	64	0.82	2.7

Table 2. Experimental conditions and results in toluene photooxidation: initial toluene concentration (HC_0), initial $FeSO_4$ seed aerosol concentration (PM_0), initial NO_x concentrations (NO_0 and $NO_{x,0}-NO_0$), ratio of PM_0/HC_0, ratio of $HC_0/NO_{x,0}$, generated SOA mass (M_0), reacted hydrocarbon (ΔHC) , and SOA yield (Y)

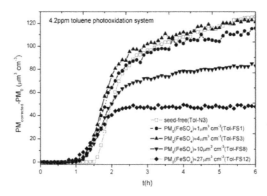

Figure 5. Temporal evolutions of SOA generation from toluene/NO_x photooxidation with different concentrations of $FeSO_4$ seed aerosols

We classified the experiments with FeSO₄ seed aerosols introduced into three groups by FeSO₄/toluene mass ratios to create SOA yield variations as a function of generated SOA mass (Figure 6). Experiments with different FeSO₄/toluene mass ratios seemed to fall into different yield curves. When FeSO₄/toluene mass ratio was lower than 4.2×10^{-4}, FeSO₄ seed aerosols had a negligible effect and SOA yields of these experiments with FeSO₄ seed aerosols coincide with the yield curve of seed-free experiments. When FeSO₄/toluene mass ratio was higher than 5.1×10^{-4}, the SOA yield curve indicated experiments with FeSO₄ seed aerosols had lower yields than seed-free experiments. Lower yield curves from the experiments with higher FeSO₄/toluene mass ratio were observed, indicating that a higher Fe/C ratio had a greater suppression effect on SOA formation from toluene/NOx photooxidation.

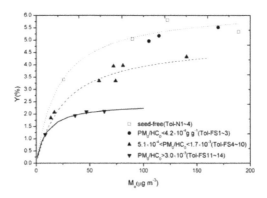

Figure 6. SOA yield (Y) variations as a function of generated SOA mass (Mo) from toluene/NOx photooxidation with/without FeSO₄ seeds

3.5. Effects of mixed (NH₄)₂SO₄ and FeSO₄ aerosols on SOA formation

Atmospheric aerosol is often a mixture of different components. We tested the effect of internal mixed (NH₄)₂SO₄ and FeSO₄ seed aerosols on SOA formation in m-xylene/NOx photooxidaiton. The experimental conditions, generated SOA mass (Mo), and SOA yield (Y) are shown in Table 3. To generate internal mixed (NH₄)₂SO₄ and FeSO₄ aerosols, a mixed solution of (NH₄)₂SO₄ and FeSO₄, in which the mass concentration ratio of (NH₄)₂SO₄ to FeSO₄ was 5:1, was used in the atomizer. So the approximately 60 μm³ cm⁻³ seed aerosols in the three experiments with mixed (NH₄)₂SO₄ and FeSO₄ seed aerosols (Xyl-FA1~3) contained about 10 μm³ cm⁻³ FeSO₄ seed aerosols and 50 μm³ cm⁻³ (NH₄)₂SO₄ seed aerosols.

As mentioned above, neither (NH₄)₂SO₄ seed aerosols nor FeSO₄ seed aerosols had obvious effects on gas phase compounds. And in the experiments in this section, we found that mixed (NH₄)₂SO₄ and FeSO₄ seed aerosols had no obvious effect on gas phase compounds either.

In Figure 7, after wall deposition correction and deduction of seed aerosols, temporal variation of particle concentrations in experiments conducted under identical initial conditions except seed aerosol concentrations (the initial concentration of m-xylene is 1.1ppm, 2.1ppm and 3.2 ppm in picture a, b and c, respectively) were compared.

Experiment No.	HC_0 ppm	PM_0 μm^3 cm^{-3}	NO_0 ppb	$NO_{x,0}-NO_0$ ppb	$HC_0/NO_{x,0}$ ppm ppm^{-1}	M_0 μg m^{-3}	ΔHC ppm	Y %
Xyl-N7	1.08	0	62	62	8.7	21	0.30	1.7
Xyl-FS1	1.01	7	58	63	8.4	8	0.29	0.7
Xyl-AS10	1.07	44	63	65	8.3	51	0.32	3.7
Xyl-FA1	1.05	62	64	69	7.9	30	0.31	2.3
Xyl-N8	2.07	0	121	120	8.6	57	0.39	3.4
Xyl-FS2	2.09	9	119	121	8.7	29	0.42	1.6
Xyl-AS11	2.15	53	121	119	9.2	119	0.52	5.4
Xyl-FA2	2.09	66	123	125	8.5	56	0.43	3.1
Xyl-N9	3.21	0	198	180	8.5	145	0.74	4.6
Xyl-FS3	3.23	11	188	182	8.8	48	0.63	1.8
Xyl-AS12	3.10	48	182	178	8.5	213	0.71	7.0
Xyl-FA3	3.16	57	179	186	8.7	117	0.74	3.7

Table 3. Experimental conditions and results in toluene photooxidation: initial toluene concentration (HC_0), initial $FeSO_4$ seed aerosol concentration (PM_0), initial NO_x concentrations (NO_0 and $NO_{x,0}-NO_0$), ratio of $HC_0/NO_{x,0}$, generated SOA mass (M_0), reacted hydrocarbon (ΔHC) and SOA yield (Y)

As indicated in Figure 7(a), comparing with the seed-free experiment Xyl-N7, both experiment Xyl-AS10 and experiment Xyl-FA1 had higher particle concentrations while experiment Xyl-FS1 had lower particle concentrations. So, in 1.1ppm m-xylene photooxidation, the presence of $(NH_4)_2SO_4$ aerosols and mixed aerosols (mixed $(NH_4)_2SO_4$ and $FeSO_4$) both increased SOA formation, while the presence of $FeSO_4$ suppressed SOA formation. In Figure 7(b) and Figure 7(c), the effects of single $(NH_4)_2SO_4$ seed aerosols (promotion effect) and single $FeSO_4$ seed aerosols (suppression effect) on SOA formation were consistent with Figure 7(a). However, the mixed aerosols seemed to have different effects on SOA formation in photooxidation systems with different initial concentrations of m-xylene. In Figure 7(b), experiment Xyl-FA2 had similar temporal variation of particle concentrations with its corresponding seed-free experiment Xyl-N8, and in Figure 7(c), experiment Xyl-FA3 had lower temporal variation of particle concentrations than its corresponding seed-free experiment Xyl-N9. It must be noted that the seed aerosols in experiments Xyl-FA1~3 had similar concentrations and components. So, aerosols at the same mixing ratio of $(NH_4)_2SO_4$ and $FeSO_4$ could either enhance or suppress SOA formation depending on the experimental conditions. It seemed that the promotion effect of $(NH_4)_2SO_4$ aerosols and the suppression effect of $FeSO_4$ aerosols competed when both of them existed. And the promotion effect of $(NH_4)_2SO_4$ aerosols was dominant with low initial hydrocarbon concentration in the competition, while the reverse was true with high initial hydrocarbon concentration. This illustrates that the interplay of different compositions of real atmosphere aerosols can lead to complex synergistic effects on SOA formation.

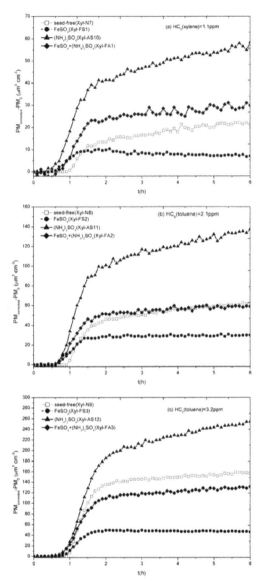

Figure 7. Temporal evolutions of generated particle concentration in experiments with/without FeSO₄, (NH₄)₂SO₄ and mixed FeSO₄ and (NH₄)₂SO₄ seed aerosols

According to the composition of the seed aerosols, experiments with inorganic seed aerosols introduced were classified into three groups. In Figure 8, SOA yield (Y) variations as a function of generated SOA mass (M_o) from m-xylene/NO_x photooxidation were plotted. The regression lines for each group were produced by fitting the data of generated SOA mass (M_o) and SOA yield (Y) into a one-product partition model. As indicated in Figure 8, experiments with the presence of $(NH_4)_2SO_4$ had a higher SOA yield curve than the seed-free experiments, while experiments with the presence of $FeSO_4$ seed aerosols had a lower one, indicating the presence of $(NH_4)_2SO_4$ and $FeSO_4$ seed aerosols increased and decreased SOA yield, respectively. For the experiments with mixed seed aerosols, their SOA yield curve was similar to or a little higher than the seed-free experiments when the SOA mass load was low, but their SOA yield curve was lower than the seed-free experiments when the SOA mass load was high.

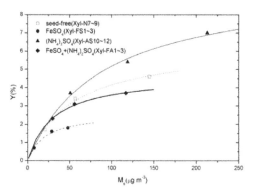

Figure 8. SOA yield (Y) variations as a function of generated SOA mass (M_o) from m-xylene/NOx photooxidation with/without $FeSO_4$, $(NH_4)_2SO_4$ and mixed $FeSO_4$ and $(NH_4)_2SO_4$ seed aerosols

3.6. Hypothesis for inorganic seed aerosols' effects

In our experiment, we observed that $FeSO_4$ seed aerosols suppressed SOA formation while $Fe_2(SO_4)_3$ seed aerosols had no effect on SOA formation. It appears that the inhibiting effect of Fe(II) involves its strong reducing properties. Hydrocarbon precursors are oxidized by OH·, $NO_3·$, etc. During the gas phase reaction, the oxidized products usually have a lower saturation vapor pressure and, as a result, condense to the aerosol phase. When these oxidized condensable compounds (CCs) containing carbonyl, hydroxyl, and carboxyl groups (Gao et al., 2004, Hamilton et al., 2005) contact ferrous iron in the aerosol phase, they may react to produce ferric iron and less condensable compounds (LCCs) or incondensable compounds (ICs). The ferrous iron may stop some CCs from being further oxidized and forming low-volatility products (Hallquist et al., 2009), including oligomers (Gao et al., 2004). The experimental results also showed that the presence of neutral $CaSO_4$ seed aerosols seed aerosols have no significant effect on photooxidation of aromatic hydrocarbons, while the presence of acid $(NH_4)_2SO_4$ seed aerosols can significantly enhance SOA generation and SOA yield. A possible mechanism is shown in Figure 9. Oligomerization is one important step during SOA formation (Nguyen et al., 2011). As proposed by (Kroll et al., 2007), the effect of $(NH_4)_2SO_4$ seed aerosols may be attributed to acid catalyzed particle-phase reactions, forming high molecular weight, low-volatility products (e.g. oligomers). These processes may deplete the semivolatile CCs in the particle phase, and enhance SOA formation by shifting the gas-particle equilibrium, which is shown in

Figure 9, and, therefore force more CCs condense to aerosol phase. Since $(NH_4)_2SO_4$ and $FeSO_4$ seed aerosols may both influence the semivolatile CCs, there is a competition for CCs to form higher-volatility products (LCCs or ICs) or low-volatility products (e.g. oligomers).

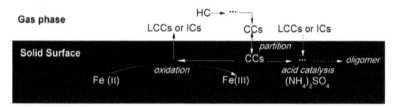

Figure 9. Hypothesized mechanism for inorganic seed aerosols' effects on SOA formation: ferrous iron Fe (II) reduces or decompose some condensable compounds (CCs), which are oligomer precursors, interrupting oligomerization and generating high volatility products (LCCs or ICs); while acid seed aerosols catalyze aerosol-phase reactions, generating oligomeric products

4. Conclusion

Effects of various inorganic seeds, including neutral inorganic seed $CaSO_4$, acidic seed $(NH_4)_2SO_4$, transition metal contained inorganic seeds $FeSO_4$ and $Fe_2(SO_4)_3$, and a mixture of $(NH_4)_2SO_4$ and $FeSO_4$, were examined during m-xylene or toluene photooxidation. Our results indicate that the presence of $CaSO_4$ seed aerosols and $Fe_2(SO_4)_3$ seed aerosols have no effect on photooxidation of aromatic hydrocarbons, while the presence of $(NH_4)_2SO_4$ seed aerosols and $FeSO_4$ seed aerosols have no effect on gas-phase reactions, but can significantly influence SOA generation and SOA yields. $(NH_4)_2SO_4$ seed aerosols enhance SOA formation and increase SOA yield due to acid catalytic effect of $(NH_4)_2SO_4$ seeds on particle-phase surface heterogeneous reactions. While $FeSO_4$ seed aerosols suppress SOA formation and decrease SOA yield possibly due to the reduction of some oligomer precursor CCs. These results reveal that many inorganic seeds are not inert during photooxidation process and can significantly influence SOA formation. These observed effects can be incorporated into air quality models to improve their accuracy in predicting SOA and fine particle concentrations.

Author details

Biwu Chu, Jingkun Jiang*, Zifeng Lu, Kun Wang, Junhua Li and Jiming Hao

State Key Laboratory of Environment Simulation and Pollution Control, School of Environment, Tsinghua University, Beijing, China

Acknowledgement

This work was supported by the National Natural Science Fundation of China (20937004, 21107060, and 21190054), Toyota Motor Corporation and Toyota Central Research and Development Laboratories Inc.

---

* Corresponding Author

5. References

Bahreini, R., Keywood, M.D., Ng, N.L., Varutbangkul, V., Gao, S., Flagan, R.C., Seinfeld, J.H., Worsnop, D.R., Jimenez, J.L., 2005. Measurements of secondary organic aerosol from oxidation of cycloalkenes, terpenes, and m-xylene using an Aerodyne aerosol mass spectrometer. Environmental Science & Technology 39, 5674-5688.

Cao, G., Jang, M., 2007. Effects of particle acidity and UV light on secondary organic aerosol formation from oxidation of aromatics in the absence of NOx. Atmospheric Environment 41, 7603-7613.

Cheng, S.-h., Yang, L.-x., Zhou, X.-h., Xue, L.-k., Gao, X.-m., Zhou, Y., Wang, W.-x., 2011. Size-fractionated water-soluble ions, situ pH and water content in aerosol on hazy days and the influences on visibility impairment in Jinan, China. Atmospheric Environment 45, 4631-4640.

Czoschke, N.M., Jang, M., Kamens, R.M., 2003. Effect of acidic seed on biogenic secondary organic aerosol growth. Atmospheric Environment 37, 4287-4299.

Duan, F.K., He, K.B., Ma, Y.L., Jia, Y.T., Yang, F.M., Lei, Y., Tanaka, S., Okuta, T., 2005. Characteristics of carbonaceous aerosols in Beijing, China. Chemosphere 60, 355-364.

Gao, S., Ng, N.L., Keywood, M., Varutbangkul, V., Bahreini, R., Nenes, A., He, J.W., Yoo, K.Y., Beauchamp, J.L., Hodyss, R.P., Flagan, R.C., Seinfeld, J.H., 2004. Particle phase acidity and oligomer formation in secondary organic aerosol. Environmental Science & Technology 38, 6582-6589.

Hallquist, M., Wenger, J.C., Baltensperger, U., Rudich, Y., Simpson, D., Claeys, M., Dommen, J., Donahue, N.M., George, C., Goldstein, A.H., Hamilton, J.F., Herrmann, H., Hoffmann, T., Iinuma, Y., Jang, M., Jenkin, M.E., Jimenez, J.L., Kiendler-Scharr, A., Maenhaut, W., McFiggans, G., Mentel, T.F., Monod, A., Prevot, A.S.H., Seinfeld, J.H., Surratt, J.D., Szmigielski, R., Wildt, J., 2009. The formation, properties and impact of secondary organic aerosol: current and emerging issues. Atmospheric Chemistry and Physics 9, 5155-5236.

Hamilton, J.F., Webb, P.J., Lewis, A.C., Reviejo, M.M., 2005. Quantifying small molecules in secondary organic aerosol formed during the photo-oxidation of toluene with hydroxyl radicals. Atmospheric Environment 39, 7263-7275.

He, K.B., Yang, F.M., Ma, Y.L., Zhang, Q., Yao, X.H., Chan, C.K., Cadle, S., Chan, T., Mulawa, P., 2001. The characteristics of PM2.5 in Beijing, China. Atmospheric Environment 35, 4959-4970.

Henry, F., Coeur-Tourneur, C., Ledoux, F., Tomas, A., Menu, D., 2008. Secondary organic aerosol formation from the gas phase reaction of hydroxyl radicals with m-, o- and p-cresol. Atmospheric Environment 42, 3035-3045.

Jang, M.S., Czoschke, N.M., Lee, S., Kamens, R.M., 2002. Heterogeneous atmospheric aerosol production by acid-catalyzed particle-phase reactions. Science 298, 814-817.

Kaiser, J., 2005. How dirty air hurts the heart. Science 307, 1858-1859.

Kroll, J.H., Chan, A.W.H., Ng, N.L., Flagan, R.C., Seinfeld, J.H., 2007. Reactions of semivolatile organics and their effects on secondary organic aerosol formation. Environmental Science & Technology 41, 3545-3550.

Lewandowski, M., Jaoui, M., Offenberg, J.H., Kleindienst, T.E., Edney, E.O., Sheesley, R.J., Schauer, J.J., 2008. Primary and secondary contributions to ambient PM in the midwestern United States. Environmental Science & Technology 42, 3303-3309.

Liggio, J., Li, S.M., 2008. Reversible and irreversible processing of biogenic olefins on acidic aerosols. Atmospheric Chemistry and Physics 8, 2039-2055.

Nguyen, T.B., Roach, P.J., Laskin, J., Laskin, A., Nizkorodov, S.A., 2011. Effect of humidity on the composition of isoprene photooxidation secondary organic aerosol. Atmospheric Chemistry and Physics 11, 6931-6944.

Odum, J.R., Hoffmann, T., Bowman, F., Collins, D., Flagan, R.C., Seinfeld, J.H., 1996. Gas/particle partitioning and secondary organic aerosol yields. Environmental Science & Technology 30, 2580-2585.

Pandis, S.N., Harley, R.A., Cass, G.R., Seinfeld, J.H., 1992. Secondary organic aerosol formation and transport. Atmospheric Environment Part a-General Topics 26, 2269-2282.

Poulain, L., Wu, Z., Petters, M.D., Wex, H., Hallbauer, E., Wehner, B., Massling, A., Kreidenweis, S.M., Stratmann, F., 2010. Towards closing the gap between hygroscopic growth and CCN activation for secondary organic aerosols - Part 3: Influence of the chemical composition on the hygroscopic properties and volatile fractions of aerosols. Atmospheric Chemistry and Physics 10, 3775-3785.

Qi, L., Nakao, S., Malloy, Q., Warren, B., Cocker, D.R., III, 2010. Can secondary organic aerosol formed in an atmospheric simulation chamber continuously age? Atmospheric Environment 44,

Satheesh, S.K., Moorthy, K.K., 2005. Radiative effects of natural aerosols: A review. Atmospheric Environment 39, 2089-2110.

Song, C., Na, K., Warren, B., Malloy, Q., Cocker, D.R., 2007. Secondary organic aerosol formation from m-xylene in the absence of NOx. Environmental Science & Technology 41, 7409-7416.

Takekawa, H., Karasawa, M., Inoue, M., Ogawa, T., Esaki, Y., 2000. Product analysis of the aerosol produced by photochemical reaction of α-pinene. Earozoru Kenkyu 15, 35-42.

Takekawa, H., Minoura, H., Yamazaki, S., 2003. Temperature dependence of secondary organic aerosol formation by photo-oxidation of hydrocarbons. Atmospheric Environment 37, 3413-3424.

Verheggen, B., Mozurkewich, M., Caffrey, P., Frick, G., Hoppel, W., Sullivan, W., 2007. alpha-Pinene oxidation in the presence of seed aerosol: Estimates of nucleation rates, growth rates, and yield. Environmental Science & Technology 41, 6046-6051.

Volkamer, R., Jimenez, J.L., San Martini, F., Dzepina, K., Zhang, Q., Salcedo, D., Molina, L.T., Worsnop, D.R., Molina, M.J., 2006. Secondary organic aerosol formation from anthropogenic air pollution: Rapid and higher than expected. Geophysical Research Letters 33, L17811.

Wu, S., Lu, Z.F., Hao, J.M., Zhao, Z., Li, J.H., Hideto, T., Hiroaki, M., Akio, Y., 2007. Construction and characterization of an atmospheric simulation smog chamber. Advances in Atmospheric Sciences 24, 250-258.

Yu, Y., Ezell, M.J., Zelenyuk, A., Imre, D., Alexander, L., Ortega, J., D'Anna, B., Harmon, C.W., Johnson, S.N., Finlayson-Pitts, B.J., 2008. Photooxidation of alpha-pinene at high relative humidity in the presence of increasing concentrations of NOx. Atmospheric Environment 42, 5044-5060.

Chapter 4

Aerosol Direct Radiative Forcing: A Review

Chul Eddy Chung

Additional information is available at the end of the chapter

1. Introduction

Aerosols affect climate in multiple ways. Aerosol absorbs or scatters radiation in the atmosphere (so-called direct effect). Aerosols, except dust, interfere mainly with solar radiation. Some aerosols act as cloud condensation nuclei (CCN), thus affecting cloud albedo and lifetime (so-called indirect effect). Dark color aerosols can be deposited on sea ice, snow packs and glaciers, thus darkening the snow and ice surfaces, and enhancing the absorption of sunlight (so-called surface darkening effect). Some of the aerosols can absorb sunlight efficiently and heat the atmosphere. This heating can burn cloud (so-called semi-direct effect). Here, I offer an overview of the aerosol direct effect on solar radiation.

The effect of aerosols on climate is normally quantified in terms of aerosol radiative forcing. Aerosol radiative forcing is defined as the effect of anthropogenic aerosols on the radiative fluxes at the top of the atmosphere (TOA) and at the surface and on the absorption of radiation within the atmosphere. The effect of the total (anthropogenic + natural) aerosols is called aerosol radiative effect or total aerosol forcing. In this chapter, I discuss various parameters that affect aerosol direct radiative effect or aerosol direct radiative forcing.

τ	AOD	Aerosol Optical Depth
τ_a	AAOD	Absorption Aerosol Optical Depth; = (1−SSA)×AOD
α	AE	Ångström Exponent for Extinction
β	AAE	Absorption Ångström Exponent
	SSA	Single Scattering Albedo
	ASY	Asymmetry parameter

Table 1. Summary of acronyms and symbols.

Aerosol direct forcing can be, and has been, estimated purely from observations alone, but the estimation has been done predominantly by a radiation model. A variety of radiation models have been used for estimating aerosol direct forcing and all of them have common input variables such as AOD (or extinction coefficient), SSA (Single Scattering Albedo), ASY (Asymmetry Parameter). These input variables have been obtained by aerosol simulation models or by aerosol observations. I review the input variables and also give an estimate of aerosol direct radiative effect.

2. Aerosol optical properties

2.1. Aerosol Optical Depth (AOD)

When a beam of light is attenuated, we call this attenuation *extinction*. Extinction is a result of scattering plus absorption. Aerosols can scatter and absorb light, and the attenuation due to aerosol is called aerosol extinction. Aerosol extinction will weaken the light intensity from I_λ to $I_\lambda + dI_\lambda$ after traversing a thickness ds in the direction of its propagation. λ represents wavelength. Then, the following equation holds:

$$d\ I_\lambda = -k_\lambda \rho I_\lambda\ ds \qquad (1)$$

where ρ is the density of the material, and k_λ denotes the mass extinction cross section (in units of area per mass). $k_\lambda\rho$ is referred to as the *aerosol extinction coefficient*, whose units are given in terms of length (typically, cm^{-1}). The aerosol extinction coefficient is the sum of the aerosol scattering coefficient and the aerosol absorption coefficient. Over the globe, the aerosol extinction coefficient is a function of space (X-Y-Z), time (T) and wavelength. The aerosol optical depth τ is a vertical integral of the aerosol extinction coefficient from the earth surface (Sfc) to the top of the atmosphere (TOA), as follows:

$$\tau_\lambda = \int_{Sfc}^{TOA} k_\lambda \rho dz \qquad (2)$$

AOD is not a function of height. AOD is the sum of AAOD (Absorption Aerosol Optical Depth) and SAOD (Scattering Aerosol Optical Depth). AAOD (τ_a) is the vertical integral of the aerosol absorption coefficient.

Another way to understand AOD is that it describes column-integrated aerosol amount in an optical sense. When aerosol mass amount is doubled, AOD should also be doubled. Aerosol mass amount, however, is not directly related to aerosol forcing, as AOD is. Furthermore, satellite observations can be used to infer AOD, not aerosol mass amount. For these collective reasons, AOD is the most fundamental variable for aerosol-climate interaction. AOD is also called AOT (Aerosol Optical Thickness). Combining Eq. 1 and Eq. 2, we find that $I_\lambda(Sfc) = I_\lambda(TOA) \times \exp(-AOD)$. Thus, aerosols with AOD of 1.0 reduce the light beam (i.e., direct radiation) by e^{-1}. e^{-1} is 0.368. In this case, the sun will appear largely hidden by aerosols at the surface. AOD of 1.0 represents a very dense aerosol layer.

AOD is a function of wavelength. The community generally uses the 550 nm value for the standard AOD. Fig. 1 shows how aerosol forcing changes with respect to AOD. When AOD is small (say, < 0.3), doubling AOD leads to doubled forcing. When AOD becomes large, added AOD translates into a smaller increase in forcing.

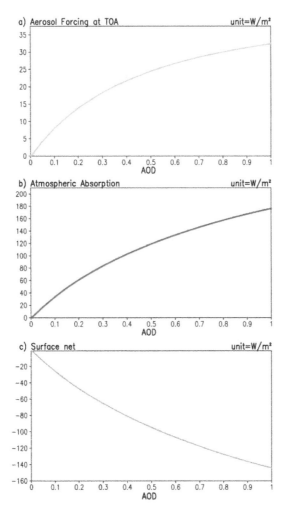

Figure 1. Annual-mean clear-sky aerosol forcing as a function of AOD at 550 nm. The simulation is made with a Monte-Carlo radiation model as in Chung et al. (2005), which only considered solar radiation. Specified parameters are SSA=0.19 at all the wavelengths, ASY=0.7 at 550 nm, α=1.4, and land surface albedo of 0.15 at a latitude of 21°N.

2.2. Single Scattering Albedo (SSA)

When photons hit an aerosol particle, some photons will be scattered while the other will be absorbed. The SSA is defined as the ratio of the scattering to the extinction. Extinction is the sum of scattering and absorption. When photons are scattered, the wavelength remains unchanged. SSA is a function of wavelength.

SSA can be computed in case of a single particle, aerosol layer or column integrated aerosols. For a single particle, the number of scattered/absorbed photons can be counted to calculate the SSA, or the scattering/extinction cross section can be measured/calculated. For a single particle, its SSA depends on particle size, particle shape and material refractive index. For an aerosol layer, the aerosol extinction/scattering coefficient can be used to compute the SSA. For column-integrated aerosols, AOD and SAOD can be used to compute the column-integrated SSA. For a group of aerosols, aerosol size distribution will affect the SSA.

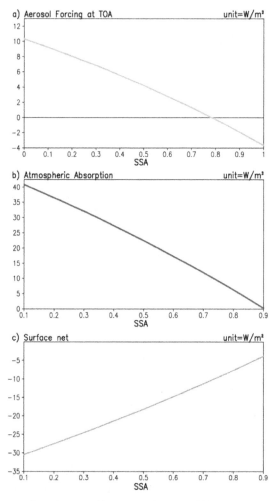

Figure 2. Annual-mean clear-sky aerosol forcing as a function of SSA. SSA is prescribed to be wavelength independent here. AOD at 550 nm is 0.1, and the rest parameters are as in Fig. 1.

Fig. 2 shows how aerosol forcing changes with respect to SSA. It is very important to note that the forcing at the TOA depends crucially on SSA. When SSA is low, the TOA forcing is positive. Conversely, the TOA forcing becomes negative with high SSAs. Aerosol forcing typically refers to the TOA forcing. Another important feature is that the surface forcing becomes larger (more negative) with lower SSA given a fixed AOD. In other words, absorbing aerosols are more effective surface dimmers.

SSA is one of the aerosol intrinsic properties. Aerosols can be classified in terms of aerosol species, the most common of which are BC (black carbon), OM (organic matter), dust, sulfate, sea salt and nitrate. Sea salt, sulfate and nitrate are known to have close to 1.0 in SSA. OM was in the past treated as 100% scattering (Myhre et al., 2007; Stier et al., 2007), but is now widely accepted to have significant absorption due to brown carbon (Andreae & Gelencsér, 2006) (BrC) component. It appears that there are large differences in the estimated magnitude of BrC absorption (Alexander et al., 2008; Chakrabarty et al., 2010; Hoffer et al., 2006). Magi (2009, 2011) analyzed air-craft data over the southern Africa and concluded that OM SSA is 0.85±0.05 at 550 nm. There is a possibility that OM SSA over the southern Africa might differ from that over other regions.

Magi (2009, 2011) also gives BC SSA. According to his field study, BC SSA is 0.19±0.05 at 550 nm. BC SSA of 0.19 is very close to 0.185 from a theoretical calculation of BC aggregates by Chung et al. (2011) and also close to 0.18 from a laboratory study by Schnaiter et al. (2005). Many studies use a very high BC SSA (typically near 0.3), and this high BC SSA results from an assumption that BC is a spherical particle. BC has a cluster structure consisting of many monomers, and Chung et al. (2011) considered the cluster structure to derive the BC SSA. When BC is assumed to be spherical, Chung et al. (2011) found BC SSA to be 0.32.

Dust SSA has been estimated to be about 0.9 by Müller et al. (2010) and 0.92 by Eck et al. (2010) at 550 nm. When dust is transported over polluted areas, non-dust particles such as BC are often attached to dust. These two field studies (Müller et al., 2010; Eck et al., 2010) probably reported polluted dust SSA. Pure dust SSA is likely to be greater than 0.92. The combination of different aerosols will determine the aerosol SSA. When aerosols are BC rich, e.g., the SSA will be low thanks to the BC component. Thus, aerosol SSA is indicative of the relative abundance of each aerosol species.

2.3. Asymmetry Parameter (ASY)

When aerosols scatter light, the phase function describes the angular distribution of scattered energy. The phase function $P(\cos\Theta)$ is a normalized function, such that

$$\int_0^{2\pi} \int_0^\pi \frac{P(\cos\Theta)}{4\pi} \sin\Theta \, d\Theta \, d\phi = 1 \qquad (3)$$

where Θ refers to the angle between the direction of incoming light and that of the scattered light. When $\Theta < \pi/2$, the scattering is called forward scattering, while the scattering is backward when $\Theta > \pi/2$.

The asymmetry parameter, or asymmetry factor, g is defined as follows:

$$g = \frac{1}{2} \int_{-1}^{1} P(\cos\Theta)\cos\Theta \, d\cos\Theta \qquad (4)$$

When the forward scattering is as much as the backward scattering, ASY becomes zero. ASY increases as the forward scattering dominates over the backward scattering. Larges particles have higher ASY. In the atmosphere, monthly-mean aerosol ASY ranges from 0.6 to 0.82 (from AERONET data analysis). AERONET (Holben et al., 2001) is a ground-based network of sun photometers located at over hundreds of stations around the world.

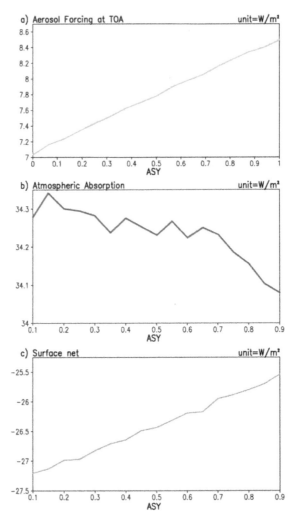

Figure 3. Annual-mean clear-sky aerosol forcing as a function of 550 nm ASY. AOD is 0.1, and the rest parameters are as in Fig. 1.

ASY has large impacts on aerosol forcing at the TOA but have little impacts on the atmospheric aerosol forcing (Fig. 3). This is because ASY does not change aerosol absorption which dominates the atmospheric forcing. Large ASYs are associated with large aerosol forcing at the TOA because aerosols with large ASY less scatter the solar radiation back to the space.

2.4. Ångström exponent (α)

α describes the wavelength dependence of AOD (or the aerosol extinction coefficient). In case of AOD Ångström exponent, the definition is as follows:

$$AOD(\lambda) = AOD(\lambda_R)(\frac{\lambda}{\lambda_R})^{-\alpha} \tag{5}$$

λ_R is the reference wavelength, and is typically 550 nm. α cannot be negative, and can be as high as 4.0. Lower α means that aerosol extinction is more independent of wavelength, which is the case for larger particles. Large particles are associated with lower α and higher ASY.

Fig. 4 shows the effect of increasing α on aerosol forcing. As α increases, the total aerosol extinction of broad-band solar radiation decreases. Thus, large α is associated with slightly less aerosol forcing.

3. Other factors controlling aerosol forcing

In Section 2, I gave an overview of aerosol optical properties and explained how these properties affect aerosol forcing. Aerosol forcing is also influenced by non-aerosol properties, notably the surface albedo and low-level cloudiness.

The surface plays an important role in case of absorbing aerosols (i.e., aerosols with low SSA). As Fig. 5 shows, higher albedo (i.e., more reflection at the surface) increases aerosol absorption and thus aerosol forcing at the TOA as well as in the atmosphere. Higher albedo increases aerosol absorption because absorbing aerosols absorb not just the downward solar radiation but also the reflected upward radiation. Higher albedo also decreases aerosol scattering back to the space, further contributing to higher aerosol forcing at TOA. Ice, snow and desert have high surface albedo.

Low-level cloud reflects solar radiation effectively, and so absorbing aerosols above low cloud have more absorption, as demonstrated by Podgorny and Ramanathan (2001). Thus, absorbing aerosols above low cloud enhance aerosol forcing, just like absorbing aerosols over reflective surfaces. The difference between low cloud and highly-reflective surface is that aerosols can be located below or above low cloud. Zarzycki and Bond (2010) studied absorbing aerosol forcing with respect to low cloud. They found that BC aerosols above low clouds explain about 20% of the global burden but 50% of the forcing.

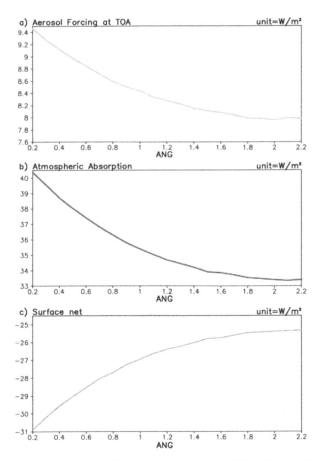

Figure 4. Annual-mean clear-sky aerosol forcing as a function of α. AOD at 550 nm is 0.1, and the rest parameters are as in Fig. 1.

4. Global distribution of aerosol optical properties

Here, I present observationally-constrained estimates of aerosol optical properties over the globe. The principal observation used here is AERONET (Aerosol Robotic NETwork), which is a ground-based network of measuring aerosol optical properties (Holben et al., 2001) as mentioned earlier. There are hundreds of AERONET sites worldwide, and all the sites are located over the land or an island. I use the monthly Level 2.0 from Version 2 product for the period 2001–2009. In this dataset, values are pre- and post-field calibrated, cloud screened and quality assured. AERONET offers AOD, SSA and ASY at multiple wavelengths. Where necessary, I logarithmically interpolated AOD and linearly interpolated SSA/ASY to the desired wavelength.

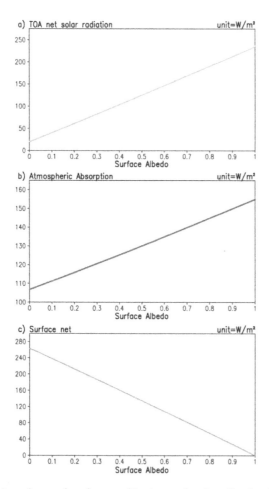

Figure 5. Figure 5. Annual-mean clear-sky aerosol forcing as a function of land surface albedo. AOD is 0.1, and the rest parameters are as in Fig. 1.

4.1. AOD at 550 nm

AERONET offers world-wide but sparsely-located AODs. To get globally gridded AOD, I use satellite observations from MODIS (MODerate resolution Imaging Spectro-radiometer) and MISR (Multi-angle Imaging Spectro-Radiometer). MODIS is a satellite sensor on board Terra satellite and Aqua satellite. I downloaded Collection 5.1 Aqua and Collection 5.0 Terra M3 AODs at 550 nm. Terra AOD and Aqua AOD on the $1^\circ \times 1^\circ$ resolution were converted to monthly combined AOD on the T42 resolution using following algorithm: If there are at least 5 values from either satellite in each T42 gridbox, a median is calculated in order to remove outliers. Then, 2001–2009 climatology for each calendar month is computed. As for

MISR AOD, I downloaded the CGAS MIL3MAE.4 product. Processing MISR AOD is similar to that of MODIS AOD.

I put together AERONET, MODIS and MISR AODs at 550 nm in the following. 1) I fill the gaps in MODIS AOD with MISR AOD using the iterative difference-successive correction method developed by Cressman (1959). MODIS does not give AOD over desert areas where MISR offers AOD. 2) The remaining gaps in MODIS+MISR AOD are filled with GOCART AOD again using Cressman(1959)'s method. 3) The spatial pattern in MODIS+MISR+GOCART AOD is coupled with the sparsely-distributed AERONET AOD values, using Chung et al. (2005)'s technique, as below.

$$N\_AODj = MMG\_AODj \times \frac{\sum_i \dfrac{AERONETj,i}{dj,i^4}}{\sum_i \dfrac{MMG\_AODj,i}{dj,i^4}} \tag{6}$$

Where N_AODj is the adjusted new value of the AOD at grid j, $AERONETj,i$ is an AERONET_ AOD at station location i nearby the grid j, dj,i is the distance between j and i, and MMG_AODj,i is the MODIS+MISR+GOCART_AOD at the grid of $AERONETj,i$. Eq. 6 is applied for each calendar month. In this assimilation method, the order of influence is AERONET > MODIS > MISR > GOCART.

Fig. 6 visualizes the assimilated AOD. AOD is large over deserts such as the Sahara and the Gobi and their downstream areas. AOD is also large over biomass burning and fossil fuel combustion areas such as East Asia, South Asia, southern Africa and Amazon.

4.2. SSA at 550 nm

To get global SSA, I put together AERONET data and GOCART simulation as follows. First, GOCART SSA is computed using GOCART AODs as follows:

$$SSA(\lambda_R) = (0.741 \times \tau_{CA}(\lambda_R) + 0.957 \times \tau_D(\lambda_R) + \tau_{rest}(\lambda_R))/\tau(\lambda_R). \tag{7}$$

CA represents carbonaceous aerosols. D represents dust. 0.957 is dust SSA. This number comes from AERONET SSA over the sites that give AAE around 2.41 ~2.42. CA SSA of 0.741 is chosen to minimize the global/annual mean difference between GOCART SSA and AERONET SSA.

Then, these GOCART SSAs are further adjusted by AERONET SSA as below.

$$\left(1-N\_SSAj\right) = \left(1-G\_SSAj\right) \times \frac{\sum_i \dfrac{1-AERONETj,i}{dj,i^4}}{\sum_i \dfrac{1-G\_SSAj,i}{dj,i^4}} \tag{8}$$

Like Eq. 6, Eq. 8 maximizes the influence of AERONET data. By applying Eq. 8, the final SSA has observational constraint on regional scales.

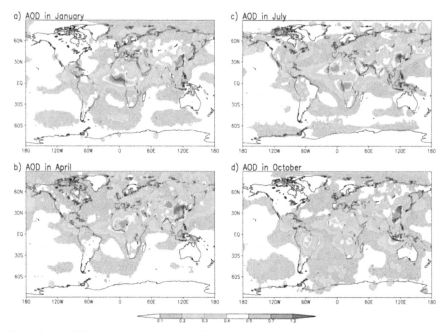

Figure 6. 2001–2009 Aerosol Optical Depth (AOD) at 550 nm, as derived by AERONET, MODIS and MISR observations.

Fig. 7 shows global aerosol SSA at 550 nm. Low SSA means absorbing aerosols. Typically, aerosols with SSA < 0.9 are considered absorbing. As the figure shows, heavy biomass burning areas such as the southern Africa show lowest SSA. This is because these areas emit large amounts of BC and relatively smaller amounts of scattering aerosols such as sulfate. Over much of the ocean, dominant aerosols are sea salt which has close to 1.0 in SSA.

4.3. ASY at 550 nm

To get global ASY, I put together AERONET data and GOCART simulation as follows. First, GOCART ASY is computed using GOCART SAODs as follows.

$$\text{ASY}(\lambda_R) = (0.62 \times \text{SAOD}_{CA}(\lambda_R) + 0.69 \times \text{SAOD}_{sul}(\lambda_R) + 0.66 \times \text{SAOD}_D(\lambda_R) \qquad (9)$$

$$+0.778 \times \text{SAOD}_{fs}(\lambda_R) + 0.85 \times \text{SAOD}_{cs}(\lambda_R))/\text{SAOD}(\lambda_R).$$

$\text{SAOD}_{sul}(\lambda_R)$ refers to SAOD at 550 nm for sulfate. "fs" refers to fine sea salt, and "cs" refers to coarse sea salt. The numbers 0.62, 0.69 and 0.66 are chosen to match AERONET ASY. The numbers 0.778 and 0.85 came from the OPAC (Optical Properties of Aerosols and Clouds) data (Hess et al., 1998). GOCART SAOD is computed from AOD and SSA, where SSA is assigned in the following: 0.19 for BC, 0.85 for OM, 0.96 for dust and 1.0 for the rest. See section 2.2 for these numbers.

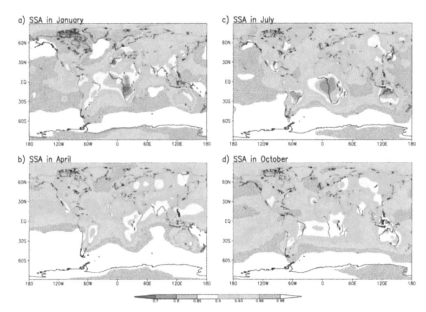

Figure 7. 2001–2009 Single Scattering Albedo (SSA) at 550 nm, as derived by AERONET and GOCART simulation.

Finally, these GOCART ASYs are adjusted by AERONET ASYs as below.

$$N\_ASYj = G\_ASYj + \dfrac{\sum_i \dfrac{AERONETi,j - G\_ASYj,i}{dj,i^4}}{\sum_i \dfrac{1}{dj,i^4}} \tag{10}$$

By applying Eq. 10, the final ASY has observational constraint on regional scales. Fig. 8 shows global ASY at 550 nm. Again, low ASY is associated with small particles. As demonstrated in Fig. 8, biomass burning areas tend to show low ASY. This is because biomass burning aerosols consist mainly of BC and OM and these two aerosol species are the smallest species. Since BC merely scatters, biomass burning aerosol ASY largely represents OM ASY. Fossil fuel combustion areas also show relatively low ASY. Large ASY values are seen over deserts and their downstream areas as well as over the ocean, because dust and sea salt are the biggest aerosols.

5. Global aerosol forcing

In section 4, I presented observationally-constrained AOD, SSA and ASY at 550 nm. In a method similar to the ASY assimilation, I generate observationally-constrained α and co-albedo Ångström exponent. Co-albedo is 1−SSA. Now, I have all the aerosol input parameters needed to compute aerosol forcing except its vertical profile. The vertical profile and the radiation model are as in Chung et al. (2005), where the Monte-Carlo Aerosol Cloud Radiation (MACR) model was adopted with the observed cloud effects from the ISCCP (International Satellite Cloud Climatology Project). All the calculations are for solar radiation and for direct effects.

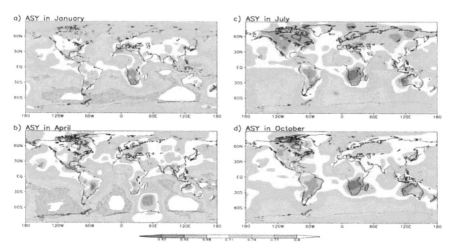

Figure 8. 2001–2009 asymmetry parameter (ASY) at 550 nm, as derived by AERONET and GOCART simulation.

Fig. 9 shows the total (natural + anthropogenic) aerosol forcing over the globe. The forcing is mostly negative, and large negative values tend to be associated with high AOD, i.e., large aerosol burden in the atmosphere. Some areas have significantly positive forcing instead. For example, the aerosols over the eastern tropical Atlantic (between Eq. and 20ºS) have huge positive forcing. This positive forcing is aided by low level cloud. To be sure, we repeated the radiation calculation without cloud (Fig. 10). The clear-sky forcing eliminates this positive-forcing feature. The remaining positive forcing in Fig. 10 is all over highly reflective surfaces such as deserts, ice. In the absence of high sulfate albedo and low cloud, the aerosol forcing is negative everywhere.

Near-zero forcing in Fig. 9 is usually associated with very little aerosol. However, the near-zero forcing that occurs between significantly positive and significantly negative forcings has a sizable amount of aerosols. Although these aerosols have near-zero forcing at the TOA, they always have large positive forcing in the atmosphere or large negative forcing at the surface. The cancellation between the surface forcing and the atmosphere forcing occurs makes the zero forcing at the TOA. This cancellation occurs when the aerosol SSA is within a certain range associated with certain surface albedo and the presence of low clouds.

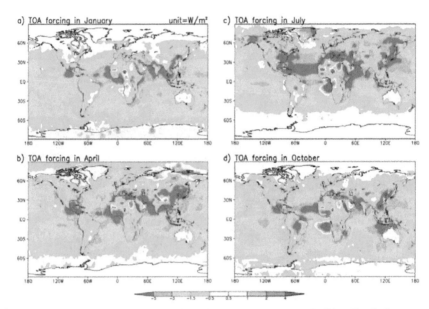

Figure 9. 2001–2009 aerosol forcing (natural + anthropogenic) estimate at the TOA. Cloud effects are included here.

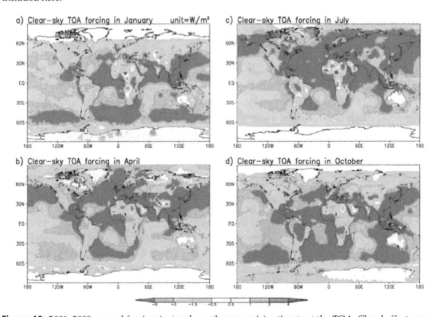

Figure 10. 2001–2009 aerosol forcing (natural + anthropogenic) estimate at the TOA. Cloud effects are not included here.

Global average aerosol forcing is summarized in Table 2. As clear in Table 2, cloud increases aerosol forcing significantly from −4.3 Wm⁻² to −2.0 Wm⁻². Surprisingly, cloud decreases the atmosphere forcing slightly, indicating that the forcing enhancement by low cloud is not as much as the forcing reduction by mid or high cloud. However, this result (i.e., cloud effects on the atmosphere forcing) is sensitive to the aerosol vertical profile, and currently there is a lot of uncertainty in aerosol vertical profile. Chung et al. (2005) used an idealized profile.

	All sky	Clear sky
TOA forcing	−2.0 Wm⁻²	−4.3 Wm⁻²
Atmosphere forcing	+4.7 Wm⁻²	+5.5 Wm⁻²
Surface forcing	−6.8 Wm⁻²	−9.7 Wm⁻²

Table 2. Global average total (natural + anthropogenic) aerosol forcing estimates.

6. Conclusion

Thus far, I have discussed fundamental aerosol optical properties and their influences on aerosol forcing, and given an observation-constrained estimate of global aerosol forcing. Although some important topics are not discussed here, the presented material here is a good starting point in studying the science of aerosol radiative forcing in my opinion.

Author details

Chul Eddy Chung

Gwangju Institute of Science and Technology, Republic of Korea

Acknowledgement

The author would like to thank Kyunghwa Lee for her technical support. This work was supported by National Research Foundation of Korea (NRF 2012-0004055).

7. References

Alexander, D. T. L., Crozier, P. A., & Anderson, J. R., (2008), Brown carbon spheres in East Asian outflow and their optical properties, *Science*, Vol. 321, No.5890, pp. 833-836.

Andreae, M. O., & Gelencsér, A., (2006), Black carbon or brown carbon? The nature of light-absorbing carbonaceous aerosols, *Atmospheric Chemistry and Physics*, Vol. 6, No. 10, pp. 3131-3148

Chakrabarty, R. K., Moosmüller, H., Chen, L. W. A., Lewis, K., Arnott, W. P., Mazzoleni, C., Dubey, M. K., Wold, C. E., Hao, W. M., & Kreidenweis, S. M., (2010), Brown carbon in tar balls from smoldering biomass combustion, *Atmospheric Chemistry and Physics*, Vol. 10, No. 13, pp. 6363-6370

Chung, C. E., Ramanathan, V., Kim, D., & Podgorny, I. A., (2005), Global anthropogenic aerosol direct forcing derived from satellite and ground-based observations, *Journal of Geophysical Research*, Vol. 110, No. D24, pp. D24207

Cressman, G. P., (1959), An operational objective analysis system, *Monthly Weather Review*, Vol. 87, No. 10, pp. 367-374.

Hess, M., Koepke, P., & I. Schult, (1998), Optical Properties of Aerosols and Clouds: The Software Package OPAC, *Bulletin of the American Meteorological Society*, Vol. 79, No. 5, pp. 831-844.

Hoffer, A., Gelencsér, A., Guyon, P., Kiss, G., Schmid, O., Frank, G. P., Artaxo, P., & Andreae, M. O., (2006), Optical properties of humic-like substances (HULIS) in biomass-burning aerosols, *Atmospheric Chemistry and Physics*, Vol. 6, No. 11, pp. 3563-3570.

Holben, B. N., Tanré, D., Smirnov, A., Eck, T. F., Slutsker, I., Abuhassan, N., Newcomb, W. W., Schafer, J. S., Chatenet, B., Lavenu, F., Kaufman, Y. J., Castle, J. V., Setzer, A., Markham, B., Clark, D., Frouin, R., Halthore, R., Karneli, A., O'Neill, N. T., Pietras, C., Pinker, R. T., Voss, K., & Zibordi, G., (2001), An emerging ground-based aerosol climatology: Aerosol optical depth from AERONET, *Journal of Geophysical Research*, Vol. 106, No. D11, pp. 12067-12097.

Magi, B. I., (2009), Chemical apportionment of southern African aerosol mass and optical depth, *Atmospheric Chemistry and Physics*, Vol. 9, No. 19, pp. 7643-7655.

Magi, B. I., (2011), Corrigendum to "Chemical apportionment of southern African aerosol mass and optical depth", *Atmospheric Chemistry and Physics*, Vol. 11, No.10, pp. 4777-4778.

Myhre, G., Bellouin, N., Berglen, T. F., Berntsen, T. K., Boucher, O., Grini, A. L. F., Isaksen, I. S. A., Johnsrud, M., Mishchenko, M. I., Stordal, F., & Tanré, D., (2007), Comparison of the radiative properties and direct radiative effect of aerosols from a global aerosol model and remote sensing data over ocean, *Tellus Series B: Chemical and Physical Meteorology*, Vol. 59, No. 1, pp. 115-129.

Podgorny, I. A., & Ramanathan, V., (2001), A modeling study of the direct effect of aerosols over the tropical Indian Ocean, *Journal of Geophysical Research*, Vol. 106, No. D20, pp. 24097-24105.

Stier, P., Seinfeld, J. H., Kinne, S., & Boucher, O., (2007), Aerosol absorption and radiative forcing, *Atmospheric Chemistry and Physics*, Vol. 7, No. 19, pp. 5237-5261.

Zarzycki, C. M., & Bond, T. C., (2010), How much can the vertical distribution of black carbon affect its global direct radiative forcing?, *Geophysical Research Letters*, Vol. 37, No. 20, pp. L20807.

Chapter 5

Production of Secondary Organic Aerosol from Multiphase Monoterpenes

Shexia Ma

Additional information is available at the end of the chapter

1. Introduction

Nonmethane volatile organic compounds (NMOCs) represent a key class of chemical species governing global tropospheric chemistry and the global carbon cycle (Fehsenfeld et al. 1992; Singh and Zimmerman 1992). The most important anthropogenic sources of hydrocarbons include fossil fuel combustion, direct release from industry, industrial processing of chemicals, and waste. The global estimated anthropogenic hydrocarbon flux is 1.0×10^{14} gC per year (Singh and Zimmerman 1992). Biological processes in both marine and terrestrial environments contribute to biogenic hydrocarbon sources. For the terrestrial biosphere, the principal hydrocarbon sources come from vegetation. In regions such as eastern North America, biogenic hydrocarbon emission rate estimates exceed anthropogenic emissions (Guenther et al. 1994). At the global scale it is estimated that vegetation emits 1.2×10^{15} gC per year, an amount equivalent to global methane emissions (Guenther et al. 1995).

Much of the recent work on emissions of biogenic volatile organic compounds (BVOCs) has focused on isoprene. However, in regions dominated by coniferous or non-isoprene emitting deciduous tree species, monoterpenes may dominate BVOC emissions. Monoterpenes comprise a significant portion of BVOC emissions (Guenther et al., 1995; Pio and Valente, 1998), and it is important to understand the atmospheric fates of monoterpenes and their oxidation products. The emission patterns of the various monoterpenes strongly depend on the type of vegetation and on the environmental conditions, however d-limonene makes up the majority of monoterpene emissions over orange groves, while α-pinene and β-pinene dominate over most other kinds of forests, especially those composed of oaks and conifers (Pio and Valente, 1998; Christensen et al., 2000). In recent years, the number of relevant studies has increased substantially, necessitating the review of this topic, including emission fluxes of monoterpenes, the effects of species and nutrient limitation on emissions, secondary organic aerosol yields via condensation and nucleation.

2. Emission fluxes of monoterpenes

2.1. Chemical structure of terpenes

Monoterpenes are organic compounds of biogenic origin whose structure may be divided into isoprene units. The more volatile mono- (C_{10}) and sesquiterpenes (C_{15}) are emitted in large quantities from the vegetation. The chemists always regard terpenes as alkenes (e.g. β-pinene, camphene), cycloalkenes (e.g. α-pinene, Δ3-carene), and dienes (e.g. β-phellandrene, α-terpinene) or as a combination of these classes (e.g. limonene, ocimene). Figure 1 shows the chemical structure of monoterpenes. Of the 14 most commonly occurring monoterpenes (α-pinene, β-pinene, Δ(3)-carene, d-limonene, camphene, myrcene, α-terpinene, β-phellandrene, sabinene, ǫ-cymene, ocimene, α-thujene, terpinolene, and γ-terpinene), the first six are usually found to be most abundant.

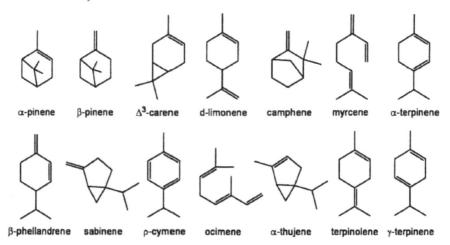

α-pinene   β-pinene   Δ³-carene   d-limonene   camphene   myrcene   α-terpinene

β-phellandrene   sabinene   ρ-cymene   ocimene   α-thujene   terpinolene   γ-terpinene

Figure 1. The chemical structures and names of monoterpenes

2.2. Biogenic sources

The biogenic volatile organic compounds (BVOCs) emitted from plants are a myriad of hydrocarbons and oxygenated and other organic compounds. These emissions occur during various stages of plant growth, plant injury and plant decay, and they are a significant source of volatile organic compounds (VOCs) in the atmosphere. Globally, emissions from BVOCs account for approximately 86% of the total of VOCs emitted while anthropogenic VOCs make up the rest (Guenther et al., 1995). BVOC emissions from trees and woody shrubs have been extensively studied (Scholes et al., 2003). The dominant compounds emitted under unperturbed conditions are isoprene, monoterpenes, sesquiterpenes and methanol. There are episodic emissions of C6 aldehydes, esters and alcohols associated with plant injuries, and episodic emissions of ethanol and acetaldehyde associated with waterlogging (Scholes et al., 2003).

Natural sources of VOC emissions to the atmosphere also include marine and fresh water, soil and sediments, microbial decomposition of organic material, geological hydrocarbon reservoirs, plant foliage and woody material. In addition, there are human influenced natural sources from harvesting or burning plant material. We have estimated emissions o f VOC only from oceans and plant foliage. VOC emissions from other sources are very uncertain but probably represent less than a few percent of total global emissions (Zimmerman, 1979; Lamb et al, 1987; Janson, 1992; Eichstaedter et al., 1992).

2.3. Emission fluxes

Estimated monoterpene emissions are dominated by α-pinene (35~70% of total) and β-pinene (15~40%). Secondary compounds of significance are d-limonene (5~20%) and β-myrcene (2~20%). Monoterpene emissions in this region are largely controlled by *Liquidambar styraciyua*, the southern pine group which includes *Pinus taeda, echinata, elliotti, palustris, virginiana,* and to a lesser extent species in the *Acer* (maple), *Magnolia, and Carya* (hickory) genera. The Pacific Coast forests and sparse coniferous forests of the Nevada Great Basin also emit primarily α-pinene. However, in eastern Montana, the western Dakotas, the Rocky Mountain Front Range of Colorado and Wyoming, the Columbia Plateau of Eastern Oregon and Washington, the Sierra Nevada Range of California, and parts of the western Great Basin, β-pinene and Δ3-carene emissions equal or exceed those of α-pinene, due to the abundance of *Pinus contorta, monticola, ponderosa,* and their subspecies. These regions and hourly summer total monoterpene flux from forests at 30°C are illustrated in Figure 2. The regions dominated by coniferous forests have the highest monoterpene flux estimates. The northern coniferous and Alaskan forests are interesting in that less than 25% of estimated monoterpene emissions are accounted for by α-pinene. The west coast, interior west, and the southern pine regions of the Piedmont and coastal southeast also have estimated fluxes which exceed 200 μg carbon m^{-2} h^{-1}. α- pinene composes over half of the emissions from these regions except for parts of the western US. Northern mixed, midwestern, Appalachian, and Ozarks forests have a higher hardwood component and have emission rates of approximately 150 μg carbon m^{-2} h^{-1}. The great plains and agricultural midwest have sparsely scattered woodlands and low emission rates (100μg carbon m^{-2} h^{-1}). These rates are rather low in comparison to those of isoprene which can exceed 10,000μg carbon m^{-2} h^{-1} from oak forests under these conditions (Geron et al., 1994; Guenther et al., 1994). Uncertainties for fluxes of these individual terpenes are difficult to estimate given analytical difficulties and gaps in knowledge of basal emission factors (EFs) and environmental/physiological controls. Current models (Guenther et al., 1994) assume ±50% uncertainty in basal EFs and roughly ±40% uncertainty in biomass and landuse accuracy (Lamb et al., 1987). An assessment of the uncertainty in monoterpene composition (MC) indicates that ±30-50% is a reasonable estimate for the six major compounds, and likely more for the eight minor compounds. This results in estimates of roughly ±150-200% for uncertainty in fluxes of individual compounds at 30°C. Canopy environment models and temperature correction algorithms can add 50-100% uncertainty to model estimates. The estimates shown here can be adjusted to ambient temperature using the exponential equations previously published (Guenther et al., 1993). However, the trees species composition and resulting flux estimates can vary significantly within these regions.

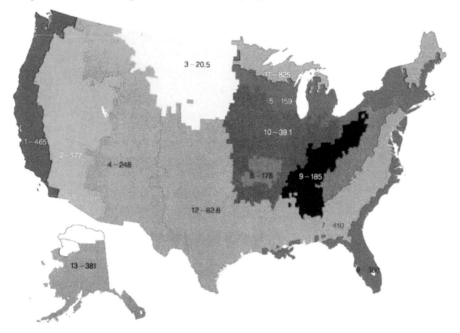

Figure 2. Regions and regional average total monoterpene flux (μg carbon $m^{-2} h^{-1}$). The numbers indicate region number follow by the average regional flux at 30°C.

2.4. Emission factors

The best-known environmental control on biogenic hydrocarbon emissions is temperature. For compounds such as monoterpenes that are released from resin ducts or glands, temperature is the dominant control factor (Figure 3). The increasing vapor pressure of these compounds with temperature explains the temperature response of emissions. In contrast, isoprene and some plant monoterpene emissions do not come from preexisting pools. In these cases the temperature response of emission is caused by the impact of temperature on the underlying metabolism (Monson et al. 1994). This effect can be described by calculating an activation energy.

Some monoterpene emissions are also dependent upon light. The light dependence is often similar to that of photosynthesis. Water and nitrogen content affect hydrocarbon emission primarily through their influences on enzyme activity of leaves. Another critical factor controlling hydrocarbon fluxes from leaves is the leaf developmental state. Monoterpene-emitting foliage shows its highest emission rates when the leaves are youngest (Lerdau 1993). This high emission rate from young leaves results from the role monoterpenes serve as defensive compounds; young leaves are most at risk from pests and pathogens so they have the highest concentrations of defensive compounds (Lerdau et al. 1994). The manner in which monoterpenes and several of the oxygenated hydrocarbons (e.g., methyl chavicol) are stored within leaves and wood leads to a large effect of tissue damage upon flux rate. These compounds are stored in specialized ducts, canals, or cavities. When herbivores partially consume tissue and expose these cavities to the atmosphere, flux rates increase by several orders of magnitude. This increase stems from the change in the resistance term in Fick's law of diffusion describing the gas flux (Lerdau 1991).

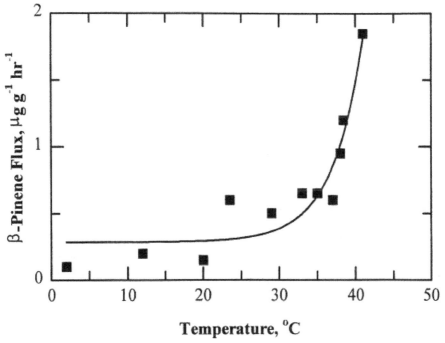

Figure 3. Variation of monoterpene emissions as function of temperature. The line represents the best fit to the data points. The emission rates are expressed per unit dry biomass weight.

2.5. Temporal and spatial distribution

Due to natural variability and analytical difficulties, monoterpene emission rates and composition are subject to considerable uncertainty. For instance, in one study 20 monoterpenes were identified using vegetation enclosures, but only eight were found in ambient air in the proximity of similar vegetation (Khalil and Rasmussen, 1992). Rough handling of vegetation during enclosure sampling has been suspected to result in artificially high emission rates (Juuti et al., 1990; Guenther et al., 1994). Summertime *Pinus taeda* monoterpene emissions were found to be 5~20 times higher following "grabbing" of foliage compared to normal experiments when rough handling was minimized (Kim et al., 1995). However, there was no effect of rough handling in the late fall. It was concluded that a film of monoterpene compounds was present on the outer surface of the foliage during summer months, but not at other times of the year. Sampling of monoterpenes from *Picea marianna* seed cones in situ versus ex situ was found to drastically affect both the emission rate as well as the monoterpene composition (Turgeon et al., 1998). Temporal and spatial factors can affect MC of a given tree species. Emissions and needle oil concentrations of $\Delta3$-carene were lower relative to other monoterpenes in the summer but were substantially higher in the spring and fall (Janson, 1993; Lerdau et al., 1995). It was concluded that the high springtime emissions of$\Delta3$-carene from *Pinus ponderosa* were due to the MC of the oleoresin in that species (Flyckt, 1979). *Picea sitchensis* (Bong.) basal emission rate (in μg carbon g^{-1} h^{-1} at leaf temperature $=30^{\circ}C$, referred to as emission factor or EF) and MC were found to change somewhat with the season (Street et al., 1996). Conversely, the oil composition of *Picea glauca* leaves and twigs remained nearly constant from summer through winter, while the oil composition of buds changed significantly during fall and winter (von Rudloff, 1972). However, it seems more typical that foliage MC remains fairly stable during most of the growing season (Hall and Langenheim, 1986; Winer et al., 1992; Street et al., 1997b; Bertin et al., 1997). This is confirmed by measurements in ambient air (Roberts et al., 1985). Diurnal variation has not been examined extensively, but was found to be negligible in *Juniperus scopulorum* (Adams and Hagerman, 1976). Likewise, little diurnal variation was found in the monoterpene composition of ambient (near canopy) air near forests in Colorado (Roberts et al., 1985), where nighttime versus daytime relative abundance of β-pinene, α-pinene, $\Delta3$-carene, camphene, and d-limonene changed by 0~4%, although nighttime total monoterpene concentration was over a factor of 2 greater due to lower dispersion. Greater seasonal and diurnal variability in light-dependent emissions of ocimene, linalool, and 1,8-cineole was found relative to the more temperature-dependent emissions of α-pinene and d-limonene, which varied little temporally from *Pinus pinea* (Staudt et al., 1997). Tree age effects have also been found for MC within this species (Adams and Hagerman, 1976) as well as for *Eucalyptus* (Street et al., 1997a) and *Picea sitchensis* (Street et al., 1996). Basal EFs were also found to decrease with tree age in these latter three studies and in *Pinus elliotti* (Kim et al., 1995). An increase in basal EFs and changes in MC of emissions following wetting of foliage have been noted (Janson, 1993; Lamb et al., 1984). Humidity has also been found to increase monoterpene emission rates at the leaf (Guenther et al., 1991) and canopy (Schade et al., 1999) levels.

3. Secondary organic aerosol formation

Most of monoterpenes are photochemically reactive in the atmosphere because of their alkenyl bonds and other properties of their structure (Seinfeld and Pandis, 1998). Two key products of the photochemical reactions of these compounds in the atmosphere are gas-phase oxidants (including ozone) and secondary organic aerosols. These gas-phase oxidants and secondary organic aerosol can have significant effects on human health when present in high concentrations in near surface air and are climatically active in the global atmosphere.

Gas-phase monoterpenes readily react with the major atmospheric oxidants such as ozone (O_3), hydroxyl radical (OH), and nitrate radical (NO_3). During the day, their concentrations are controlled by OH and O_3, and at night they are controlled by NO_3, with monoterpene lifetimes on the order of a few hours in both cased. Regardless of the initial oxidant, gas-phase oxidation of monoterpenes results in a wide variety of polyfunctional carboxylic acids, ketones, aldehydes, peroxides, and alcohols (Kavouras et al., 1998; Yu et al., 1999; Claeys et al., 2007; Presto et al., 2005; Gao et al., 2004; Dalton et al., 2005; Glasius et al., 1999; Hakola et al., 1994; Jaoui et al., 2005; Leungsakul et al., 2005). Many of these species have sufficiently low vapor pressure to partition into pre-existing particulate matter. In addition, monoterpenes can partition into aqueous particles or cloud droplets by wet deposition and undergo oxidation via aqueous chemistry, with droplets subsequently drying out into organic particles.

SOAs are formed by chemical reactions from the condensation and gas/particle partitioning of semivolatile products of hydrocarbon oxidation by OH, O_3 and NO_3. It is currently believed that approximately 85-90% of SOAs are produced by oxidation of biogenic emissions, which corresponds to 12.2 Tg SOA/yr and 1150 TgC/yr (Kanakidou et al., 2005; Stavrakou et al.,2009; Tsigaridis et al., 2007). Some authors suggest that 91% of SOAs are considered to be caused by O_3 and OH oxidation (Chung et al., 2002), while others consider that ozonolysis dominates in SOA production compared to OH and NOx oxidation (Kanakidou et al., 2005). 78% of SOAs are specifically products of monoterpene oxidation (Chung et al., 2002). Monoterpenes α-pinene and b-pinene are expected to be the major sources of SOAs (Kleindienst et al., 2007; Yu et al., 1999).

3.1. Chamber experiments

A number of laboratory experiments have been carried out to study SOA formation by oxidation of monoterpenes (e.g. α-pinene, β-pinene, d-limonene, Δ3-carene, sabinene) with NO_x, O_3 and OH (Lee et al., 2004; Jonsson et al., 2008; Northcross and Jang, 2007; Czoschke et al., 2003; Spittler et al., 2003; Colville et al., 2004; Seinfeld et al., 2001). The formation of SOA in the presence of NOx has been studied for several monoterpenes: α-pinene, β-pinene, limonene, Δ3-carene, myrcene, β-ocimene, sabinene, α-terpinene, α-terpinolene (Pandis et al., 1991; Hoffmann et al., 1997; Griffin et al., 1999; Klinedinst and Currie, 1999; Presto et al., 2005; Lee et al., 2006; Zhang et al., 2006). The complexity of the dependence of NOx concentration on monoterpene SOA yields was first noted by Pandis et al. (1991) who found that the SOA yield from the photooxidation of α-pinene increased with increasing VOC/NOx ratio and maximized at 8% for 10–20 ppbC:ppb NOx. The SOA yield was found to decrease for VOC/NOx ratios greater than 20. More recently, Presto et al. (2005) showed that SOA yields from ozonolysis of α-pinene are lower under high- NOx (<15 ppbC:ppb NOx) than low-NOx conditions. The reduction of the SOA yields was attributed to the formation of higher volatility products under high NOx, including evidence of nitrate-containing organic species. The NOx dependence becomes increasingly complex when the monoterpene consists of two double bonds with differing reactivity, as is the case for limonene. Zhang et al. (2006) indicated that two competing effects contribute to the NOx dependence of the SOA yield: reactive uptake by the aerosol and gas-phase oxidation. In the presence of ozone and at low NOx, heterogeneous ozonolysis of the exo double bond generates condensed-phase secondary generation oxidation products. However, at high NOx, gas-phase oxidation of the double bond becomes the dominant process for aerosol product generation.The yield dependence on NOx is complicated when oxidation by ozone or hydroxyl radicals takes place in the presence of NOx. The primary reason is thought to be that under high NOx conditions, organo-peroxy radicals (RO$_2$) react with NO and NO$_2$ instead of with peroxy radicals (RO$_2$ or HO$_2$). A number of different critical values (VOC/NOx) have been suggested (e.g. 10–15:1, Pandis et al., 1991, 8:1, Presto et al., 2005, 3–10:1, Lane et al., 2008) for the point of 50:50 branching between the two reaction paths. As the OH/O$_3$ ratio also depends on the VOC/NOx ratio, it is difficult to separate changes in yield due to changes in concentrations of OH and O$_3$ from changes in product distribution due to the presence of NOx (Presto et al., 2005). However, the approach adopted by Lane et al. (2008) is to identify product yields for each fundamental pathway (i.e., high-NOx dominated by RO$_2$ + NO, low-NOx by RO$_2$ +HO$_2$) and then to assume that SOA yields can be calculated by a linear combination of the "pure" mass yields scaled by the strength of each reaction pathway. Clearly, anthropogenic processes that perturb these branching ratios from their background values will influence overall SOA formation.

3.2. Field studies

The high abundant vegetation in Amazonia make this region a global hotspot for the emission of various biogenic volatile organic compounds (e.g. Keller et al., 2009). Previously it was thought that the vast emissions of VOC in the Amazon area would deplete the oxidative potential of the atmosphere (Lelieveld et al., 2002) and thus constrain the BSOA production. However, there have been strong indications that HOx chemistry remains substantially more vigorous under low-NOx conditions than standard mechanisms predict (Thornton et al., 2002). This was confirmed by a more recent study by Lelieveld et al. (2008) showing that the oxidative potential in the pristine rainforest is maintained through the recycling of OH via organic peroxy radical reactions under low NOx conditions. Most attempts to estimate Amazonian BSOA production (e.g. Penner et al., 2001; Kanakidou et al., 2005) with global models have considered only monoterpenes as precursors. More recent efforts (e.g. Heald et al., 2008) have included isoprene, which has been shown to be a significant SOA precursor (Claeys et al., 2004), but still omit some potentially significant precursors such as the highly reactive sesquiterpenes which have very high BSOA yields. Both observational and modelling studies have concluded that 80% of the aerosol mass in the Amazon forest is of biogenic origin (Artaxo et al., 1990; Artaxo and Hansson, 1995; Heald et al., 2008; Chen et al., 2009) either due to primary or secondary sources (Kanakidou et al., 2005), however there is a large uncertainty in this number. One has to keep in mind that unless we have an accurate knowledge of the emissions of primary aerosols, it will be difficult to assess the relative importance of the SOA. Measurements made during the Amazonian Aerosol Characterization Experiment 2008 (AMAZE-08), under near pristine conditions, show that the submicrometer aerosol is mostly composed of BSOA, while supermicron particles consist mainly of primary organic matter (Pöschl et al., 2010). While the submicrometer fraction was found to account for more than 99% of the ca. 200 cm^{-3} particles observed, approximately 70% of the ca. 2μg m^{-3} total particle mass was made up by the supermicron particles. The ongoing development of the Amazon is resulting in an enhanced anthropogenic influence on regional atmospheric composition. Changes in anthropogenic pollutants (particularly NOx and SO_2) can alter the SOA formation potential (Kroll et al., 2005; Ng et al., 2007) and the relative roles of different oxidation pathways. It is typically assumed that more SOA is formed from monoterpenes under low-NOx conditions, such as prevail in the Amazon. Therefore a change in the amount of NOx due to anthropogenic emissions would lead to a change in the amount of SOA. The magnitude and the sign of this change depend on the predominant VOC species.

3.3. Modeling

Current BVOC emission and air quality models aggregate all monoterpene ($C_{10}H_{16}$) compounds, assuming that their fate in the atmosphere is similar. However, studies have shown that individual monoterpene compounds may react quite differently (Hoffmann et al., 1997; Atkinson et al., 1992; Atkinson, 1990; Yokouchi and Ambe, 1985). Reaction rates with O_3, OH, and NO_3 radicals can vary by an order of magnitude between these compounds. Aerosol yields can likewise vary significantly. Monoterpenes with exocyclic double bonds, such as b-pinene and sabinene, tend to form more aerosols following ozonolysis compared to those with endocyclic double bonds, such as α-pinene and Δ^3-carene (Hatakeyama et al., 1989). Those with two double bonds can react to produce even higher aerosol yields, depending on the vapor pressure of the reaction products. Open-chain monoterpenes, such as myrcene, linalool, and ocimene (Hoffmann et al., 1997), tend to produce lower aerosol yields under most circumstances. It was recently concluded that it is not possible to use generalized descriptions of terpene chemistry in models (Hallquist et al., 1999). Aerosol forming potentials of terpenes discussed here could be partially explained by their structural characteristics (Griffin et al., 1999). It was concluded that most biogenic hydrocarbons would have to be accounted for individually when modeling atmospheric aerosol formation.

Author details

Shexia Ma
Center for Research on Urban Environment,
South China Institute of Environmental Sciences (SCIES),
Ministry of Environmental Protection (MEP), Guangzhou, China

4. References

Artaxo, P. and Hansson, H. C.: Size Distribution of Biogenic Aerosol-Particles from the Amazon Basin, Atmos. Environ., 29, 393–402, 1995.

Artaxo, P., Maenhaut, W., Storms, H., and Vangrieken, R.: Aerosol Characteristics and Sources for the Amazon Basin during theWet Season, J. Geophys. Res.-Atmos., 95, 16971–16985, 1990.

Atkinson, R., 1990. Gas-phase tropospheric chemistry of organic compounds: a review. Atmospheric Environment 24A (1), 1-41.

Atkinson, R., Aschmann, S.M., Arey, J., Shorees, B., 1992. Formation of OH radicals in the gas phase reactions of O3 with a series of terpenes. Journal of Geophysical Research 97 (D5), 6065-6073.

Adams, R.P., Hagerman, A., 1976. A comparison of the volatile oils of mature versus young leaves of *Juniperus scopulorum*: chemosystematic signi"cance. Biochemical Systematics and Ecology 4, 75-79.

Bertin, N., Staudt, M., Hansen, U., Seufert, G., Ciccioli, P., Foster, P., Fugit, J.L., Torres, L., 1997. Diurnal and seasonal course of monoterpene emissions from *Quercus ilex* (L.) under natural conditions application of light and temperature algorithms. Atmospheric Environment 31 (S1), 135-144.

Christensen, C.S., Hummelshoj, P., Jensen, N.O., Larsen, B., et al., Determination of the terpene flux from orange species and Norway spruce by relaxed eddy accumulation. Atm. Environ. 2000, 34, 3057-3067.

Chen, Q., Farmer, D. K., Schneider, J., Zorn, S. R., Heald, C. L., Karl, T. G., Guenther, A., Allan, J. D., Robinson, N., Coe, H., Kimmel, J. R., Pauliquevis, T., Borrmann, S., Poschl, U., Andreae, M. O., Artaxo, P., Jimenez, J. L., and Martin, S. T.: Mass spectral characterization of submicron biogenic organic particles in the Amazon Basin, Geophys. Res. Lett., 36, L20806, doi:10.1029/2009gl039880, 2009.

Chung, S.H. & Seinfeld, J.H., Global distribution and climate naceo forcing of carbous aerosols. *J. Geophys. Res.* 107 (2002).

Claeys, M., Graham, B., Vas, G.,Wang,W., Vermeylen, R., Pashynska, V., Cafmeyer, J., Guyon, P., Andreae, M. O., Artaxo, P., and Maenhaut, W.: Formation of secondary organic aerosols through photooxidation of isoprene, Science, 303, 1173–1176, 2004

Claeys, M., Szmigielski, R., Kourtchev, I., Van der Veken, P. et al., Hydroxydicarboxylic acids: Markers for secondary organic aerosol from the photooxidation of a-pinene, Environ. Sci. Technol, 2007, 41, 1628-1634

Czoschke, N.M., Jang, M., and Kamens, R.M., Effect of acidic seed on secondary biogenic organic aerosol growth. *Atmospheric Environment* 37 (30), 4287-4299 (2003).

Colville, C.J. and Griffin, R.J.R.J., The roles of individual oxidants in secondary organic aerosol formation from Δ3- carene: 1. gas-phase hemical mechanism. *Atmospheric Environment* 38 (24), 4001-4012 (2004).

Dalton, C.N., Jaoui, M., Kamens, R.M., Glish, G.L., Continuous real-time analysis of products from the reaction of some monoterpenes with ozone using atmospheric sampling glow discharge ionization coupled to a quadrupole ion trap mass spectrometer. Anal. Chem. 2005, 77, 3156-3163.

Eichstaedter, G., W. Schuermann, R. Steinbrecher and H. Ziegler, Diurnal cycles of soil and needle monoterpene emission rates and simultaneous gradient measurements of monoterpene concentrations in the stem region and above a Norway spruce canopy, EUROTRAC symposium '92, The Hague, Netherlands, 1992.

Fehsenfeld, F. C., and Coauthors, 1992: Emissions of volatile organic compounds from vegetation and the implications for atmospheric chemistry. *Global Biogeochem. Cycles,* 6, 389–430.

Flyckt, D.L., 1979. Seasonal variation in the volatile hydrocarbon emissions from ponderosa pine and red oak. Master's Thesis, Washington State University, pp. 1-52.

Guenther, A., Monson, R.K., Fall, R., 1991. Isoprene and monoterpene emission rate variability: Observations with eucalyptus and emission rate algorithm development. Journal of Geophysical Research 96(D6), 10,799-10,808.

Geron, C., Guenther, A., Pierce, T., 1994. An improved model for estimating emissions of volatile organic compounds from forests in the eastern United States. Journal of Geophysical Research 99(D6), 12,773-12,791.

Griffin, R.J., Cocker III, D.R., Flagan, R.C., Seinfeld, J.H., 1999. Biogenic aerosol formation from the oxidation of biogenic hydrocarbons. Journal of Geophysical Research 104 (D3), 3555-3567.

Guenther, A., Zimmerman, P., Harley, P., 1993. Isoprene and monoterpene emission rate variability: models evaluations and sensitivity analyses. Journal of Geophysical Research 98(D7), 12,609-12,617.

Guenther, A., Zimmerman, P. and M. Wildermuth, 1994: Natural volatile organic compound emission rate emissions for U.S. woodland landscapes. *Atmos. Environ.,* 28, 1197–1210.

Guenther, A. and Coauthors, 1995: A global model of natural volatile organic compound emissions. *J. Geophys. Res.,* 100, 8873–8892.

Gao, S., Keywood, M., Ng, N.L., et al., Low-molecular-weight and oligomeric components in secondary organic aerosol from the ozonolysis of cycloalkenes and a-pinenes. J.Phys.Chem. A 2004,108,10147-10164

Glasius, M., Duane, M., Larsen, B.R., Determination of polar terpene oxidation products in aerosol by liquid chromatography-ion trap mass spectrometry, J.Chromatogr. A 1999, 833,121-135

Griffin, R. J., Cocker, D. R., Flagan, R. C., and Seinfeld, J. H.: Organic aerosol formation from the oxidation of biogenic hydrocarbons, J. Geophys. Res.-Atmos., 104, 3555–3567, 1999.

Heald, C. L., Henze, D. K., Horowitz, L. W., Feddema, J., Lamarque, J. F., Guenther, A., Hess, P. G., Vitt, F., Seinfeld, J. H., Goldstein, A. H., and Fung, I.: Predicted change in global secondary organic aerosol concentrations in response to future climate, emissions, and land use change, J. Geophys. Res.-Atmos., 113, D05211, doi:10.1029/2007jd009092, 2008.

Hoffmann, T., Odum, J. R., Bowman, F., Collins, D., Klockow, D., Flagan, R. C., and Seinfeld, J. H.: Formation of organic aerosols from the oxidation of biogenic hydrocarbons, J. Atmos. Chem., 26, 189–222, 1997.

Hall, G.D., Langenheim, J.H., 1986. Temporal changes in the leaf monoterpenes of *Sequoia sempervirens*. Biochemical Systematics and Ecology 14 (1), 61-69.

Hakola, H., Arey, J., Aschmann, S.M., Atkinson, R., Product formation from the gas-phase reactions of OH Radicals and O3 with a series of monoterpenes, J.Atm. Chem, 1994, 18, 75-102

Hallquist, M., Wangberg, I., Ljungstrom, E., Barnes, I., Becker, K., 1999. Aerosol and product yields from NO3 radical initiated oxidation of selected monoterpenes. Environmental Science and Technology 33 (4), 553-559.

Hatakeyama, S., Izumi, K., Fukuyama, T., Akimoto, H., 1989. Reactions of ozone with a-pinene and b-pinene in air: yields of gaseous and particulate products. Journal of Geophysical Research 94(D10), 13,013-13,024.

Hoffmann, T., Odum, J. R., Bowman, F., Collins, D., Klockow, D., Flagan, R. C., and Seinfeld, J. H.: Formation of organic aerosols from the oxidation of biogenic hydrocarbons, J. Atmos. Chem., 26, 189–222, 1997.

Jaoui, M., Kleindienst, T.E., Lewandowski, M., et al., Identification and quantification of aerosol polar oxygenated compounds bearing carboxylic or hydroxyl groups. 2. organic tracer compounds from monoterpenes. Environ. Sci. Technol, 2005, 39, 5661-5673

Janson, R., Monoterpenes from the boreal coniferous forest, PhD thesis, University of Stockholm, 1992.

Janson, R.W., 1993. Monoterpene emissions from Scots pine and Norwegian spruce. Journal of Geophysical Research 98, 2839-2850.

Juuti, S., Arey, J., Atkinson, R., 1990. Monoterpene emission rate measurements from a Monterey pine. Journal of Geophysical Research 95 (D6), 7515-7519.

Jonsson, Å.M., Hallquist, M., & Ljungström, E., The effect of temperature and water on secondary organic aerosol formation from ozonolysis of limonene, Δ3-carene and α-pinene. *Atmos. Chem. Phys.* 8 (21), 6541-6549 (2008).

Janson, R.W., 1993. Monoterpene emissions from Scots pine and Norwegian spruce. Journal of Geophysical Research 98, 2839-2850.

Kanakidou, M. et al., Organic aerosol and global climate review modelling: a *.Atmos. Chem. Phys.*5 (4), 1053-1123 (2005).

Kanakidou, M., Seinfeld, J. H., Pandis, S. N., Barnes, I., Dentener, F. J., Facchini, M. C., Van Dingenen, R., Ervens, B., Nenes, A., Nielsen, C. J., Swietlicki, E., Putaud, J. P., Balkanski, Y., Fuzzi, S., Horth, J., Moortgat, G. K., Winterhalter, R., Myhre, C. E. L., Tsigaridis, K., Vignati, E., Stephanou, E. G., and Wilson, J.: Organic aerosol and global climate modelling: a review, Atmos.Chem. Phys., 5, 1053–1123, doi:10.5194/acp-5-1053-2005, 2005.

Keller, M., Bustamante, M., Gash, J., and Silva Dias, P. (Eds.): Geophysical Monograph Series, Volume 186, 576 pp., hardbound, ISBN 978-0-87590-476-4, AGU Code GM1864764, 2009.

Klinedinst, D. B. and Currie, L. A.: Direct quantification of PM2.5 fossil and biomass carbon within the Northern Front Range Air Quality Study's domain, Environ. Sci. Technol., 33, 4146–4154,1999.

Kleindienst, T.E. *et al.*, Estimates of the contributions of biogenic and anthropogenic hydrocarbons to secondary aerosol organic at a southeastern US location. *Atmospheric Environment* 41 (37), 8288-8300 (2007).

Kavouras, I.G., Mihalopoulos, N., Stephanou, E.G. Formation of atmospheric particles from organic acids produced by forests. Nature 1998, 395,683-686

Kroll, J. H., Ng, N. L., Murphy, S. M., Flagan, R. C., and Seinfeld, J. H.: Secondary organic aerosol formation from isoprene photooxidation under high-NOx conditions, Geophys. Res. Lett., 32, L18808, doi:10.1029/2005gl023637, 2005.

Khalil, M.A., Rasmussen, R.A., 1992. Forest hydrocarbon emissions: relationships between fluxes and ambient concentrations. Journal of Air and Waste Management Association 42 (6), 810-813.

Kim, J.-C., Allen, E.R., Johnson, J.D., 1995. Terpene emissions in a southeastern pine forest. Air and Waste Management Association 88th Annual Mtg and Ex, No. 95-WA74A.03, pp. 2-16.

Lane, T. E., Donahue, N. M., and Pandis, S. N.: Effect of NOx on secondary organic aerosol concentrations, Environ. Sci. Technol., 42, 6022–6027, 2008.

Lee, A., Goldstein, A. H., Kroll, J. H., Ng, N. L., Varutbangkul, V., Flagan, R. C., and Seinfeld, J. H.: Gas-phase products and secondary aerosol yields from the photooxidation of 16 different terpenes, J. Geophys. Res.-Atmos., 111, D17305, doi:10.1029/2006jd007050, 2006.

Lee, S., Jang, M., Kamens, R.M., SOA formation from the photooxidation of α-pinene in the presence of freshly emitted diesel soot exhaust. *Atmospheric Environment* 38 (16), 2597-2605 (2004).

Leungsakul, S., Jeffries, H.E., Karmens, R.M., A kinetic mechanism for predicting secondary organic aerosol formation from the reactions of D-Limonene in the presence of oxides of nitrogen and natural sunlight. Atm. Environ. 2005, 39, 7063-7082

Lelieveld, J., Peters, W., Dentener, F. J., and Krol, M. C.: Stability of tropospheric hydroxyl chemistry, J. Geophys. Res.-Atmos., 107, 4715, doi:10.1029/2002jd002272, 2002.

Lelieveld, J., Butler, T. M., Crowley, J. N., Dillon, T. J., Fischer, H., Ganzeveld, L., Harder, H., Lawrence, M. G., Martinez, M., Taraborrelli, D., and Williams, J.: Atmospheric oxidation capacity sustained by a tropical forest, Nature, 452, 737–740, 2008.

Lerdau, M., 1991: Plant function and biogenic terpene emissions. *Trace Gas Emissions from Plants*, T. Sharkey, E. Holland, and H. Mooney, Eds., Academic Press, 121–134.

Lerdau, M., 1993: Ecological controls over monoterpene emissions from conifers. Ph.D. thesis, Stanford University, 108 pp. [Available from University Microfilms Dissertation Services, http://www.umi.com/hp/Support/DServices/.]

Lerdau, M., S. Dilts, H. Westberg, B. Lamb, and G. Allwine, 1994: Monoterpene emission from ponderosa pine. *J. Geophys. Res.*, 99, 16 609–16 615.

Lamb, B., Westberg, H., Quarles, T., Flyckt, D.L., 1984. Natural hydrocarbon emission rate measurements from selected forest sites. US Environmental Protection Agency. EPA-600/3-84-001, NTIS No. PB84-124981.

Lamb, B., Guenther, A., Gay, D., Westberg, H., 1987. A national inventory of biogenic hydrocarbon emissions. Atmospheric Environment 21 (8), 1695-1705.

Lerdau, M., Matson, P., Fall, R., Monson, R.K., 1995. Ecological controls over monoterpene emissions from Douglas fir (*Pseudotsuga menziesii*). Ecology 76 (8), 2640-2647.

Monson, R. K., P. C. Harley, M. E. Litvak, M. Wildermuth, A. B. Guenther, P. R. Zimmerman, and R. Fall, 1994: Environmental and developmental controls over the seasonal pattern of isoprene emission from aspen leaves. *Oecologia*, 99, 260–270

Northcross, A.L. and Jang, M., Heterogeneous SOA yield from ozonolysis of monoterpenes in the presence of inorganic acid. *Atmospheric Environment* 41 (7), 1483-1493 (2007).

Ng, N. L., Kroll, J. H., Chan, A. W. H., Chhabra, P. S., Flagan, R. C., and Seinfeld, J. H.: Secondary organic aerosol formation from m-xylene, toluene, and benzene, Atmos. Chem. Phys., 7, 3909–3922, doi:10.5194/acp-7-3909-2007, 2007.

Pandis, S. N., Paulson, S. E., Seinfeld, J. H., and Flagan, R. C.: Aerosol Formation in the Photooxidation of Isoprene and Beta- Pinene, Atmos. Environ. Part a, 25, 997–1008, 1991.

Pöschl, U., Martin, S. T., Sinha, B., Chen, Q., Gunthe, S. S., Huffman, J. A., Borrmann, S., Farmer, D. K., Garland, R. M., Helas, G., Jimenez, J. L., King, S. M., Manzi, A., Mikhailov,

E., Pauliquevis, T., Petters, M. D., Prenni, A. J., Roldin, P., Rose, D., Schneider, J., Su, H., Zorn, S. R., Artaxo, P., and Andreae, M. O.: Rainforest Aerosols as Biogenic Nuclei of Clouds and Precipitation in the Amazon, Science, 329, 1513–1516, doi:10.1126/science.1191056, 2010.

Presto, A. A., Hartz, K. E. H., and Donahue, N. M.: Secondary organic aerosol production from terpene ozonolysis. 2. Effect of NOx concentration, Environ. Sci. Technol., 39, 7046-7054, 2005.

Penner, J. E., Hegg, D., and Leaitch, R.: Unraveling the role of aerosols in climate change, Environ. Sci. Technol., 35, 332A–340A, 2001.

Presto, A.A., Huff Hartz, K.E., Donahue, N.M., Secondary organic aerosol production from terpene ozonolysis. 1. Effect of UV Radiation. Environ. Sci. Technol, 2005, 39, 7036-7045

Pio, C., Valente, A. A. Atmospheric fluxes and concentrations of monoterpenes in resin-tapped pine forests. Atm. Environ. 1998, 32, 683-691.

Roberts, J.M., Hahn, C.J., Fehsenfeld, F.C., Warnock, J.M., Albritton, D.L., Sievers, R.E., 1985. Monoterpene hydrocarbons in the nighttime troposphere. Environmental Science and Technology 19 (4), 364-369

Seinfeld, J.H., Erdakos, G.B., Asher, W.E., and Pankow, J.F., Modeling the Formation of Secondary Organic Aerosol (SOA). 2. The Predicted Effects of Relative Humidity on Aerosol Formation in the α-Pinene, β-Pinene, Sabinene, Δ3-Carene, and Cyclohexene Ozone Systems. *Environmental Science & Technology* 35 (15), 3272-3272 (2001)

Spittler, M. et al., Reactions of NO3 radicals with limonene and [alpha]-pinene: Product and SOA formation. *Atmospheric Environment* 40 (Supplement 1), 116-127 (2006).

Stavrakou, T. et al., Evaluating the performance of pyrogenic and biogenic emission inventories against one decade of space-based formaldehyde columns. *Atmos. Chem. Phys.* 9 (3), 1037-1060 (2009).

Staudt, M., Bertin, N., Hansen, U., Seufert, G., Ciccioli, P., Foster, P., Frenzel, B., Fugit, J.L., 1997. Seasonal and diurnal patterns of monoterpene emissions from *Pinus pinea* (L.) under field conditions. Atmospheric Environment 31 (S1),145-156.

Schade, G.W., Goldstein, A.H. Lamanna, M.S., 1999. Are monoterpene emissions influenced by humidity? Geophysical Research Letters 26(14), 2187-2190.

Street, R.A., Duckham, S.C., Hewitt, C.N., 1996. Laboratory and field studies of biogenic volatile organic compound emissions from Sitka spruce (*Picea sitchensis* Bong.) in the United Kingdom. Journal of Geophysical Research 101(D17), 22,799-22,806.

Street, R.A., Hewitt, C.N., Mennicken, S., 1997a. Isoprene and monoterpene emissions from a Eucalyptus plantation in Portugal. Journal of Geophysical Research 102(D13), 15,875-15,887.

Street, R.A., Owen, S., Duckham, S.C., Boissard, C., Hewitt, C.N., 1997b. E!ect of habitat and age on variations in volatile organic compound (VOC) emissions from *Quercus ilex and Pinus pinea*. Atmospheric Environment 31 (S1), 89-100.

Singh, H. B., and P. Zimmerman, 1992: Atmospheric distribution and sources of nonmethane hydrocarbons. *Gasesous Pollutants: Characterization and Cycling,* J. O. Nriagu, Ed., John Wiley and Sons, 235 pp.

Scholes, M.C., Matrai, P.A., Andreae, M.O., Smith, K.A., Manning, M.R., Artaxo, P., Barrie, L.A., Bates, S.S., Butler, J.H., Ciccioli, P., Cieslik, S.A., Delmas, R.J., Dentener, F.J., Ducce, R.A., Erickson, D.J., Galbally, I.E., Guenter, A.B., Jaenicke, R., Ja¨ jne, A.J., Kienee, R.P., Lacaux, J.P., Liss, P.S., Malin, G., Matsonm, P.A., Mosier, A.R., Neue, H.U., Paerl, H.W., Platt, U.F., Quinn, P.K., Seiler, W., Weiss, R.F., 2003. Biosphere–atmosphere interactions. In: Brasseur, G.P., Prinn, R.G., Pszenny, A.A.P. (Eds.), Atmospheric Chemistry in a Changing World: An Integration and Synthesis of a Decade of Tropospheric Chemistry Research: The International Global Atmospheric Chemistry Project of the International Geosphere-Biosphere Programme. Springer, New York, pp. 19–71.

Turgeon, J.J., Brockerhoff, E.G., Lombardo, D.A., MacDonald, L., Grant, G.G., 1998. Differences in composition and release rate of volatiles emitted by black spruce seed cones sampled in situ versus ex situ. Canadian Journal of Forest Research 28, 311-316.

Tsigaridis, K. & Kanakidou, M., Secondary organic aerosol importance in the future atmosphere. *Atmospheric Environment* 41 (22), 4682-4692 (2007).

Thornton, J. A., Wooldridge, P. J., Cohen, R. C., Martinez, M., Harder, H., Brune,W. H.,Williams, E. J., Roberts, J. M., Fehsenfeld, F. C., Hall, S. ., Shetter, R. E., Wert, B. P., and Fried, A.: Ozone production rates as a function of NOx abundances and HOx production rates in the Nashville urban plume, J. Geophys. Res.-Atmos., 107, 4146, doi:10.1029/2001jd000932, 2002.

von Rudloff, E., 1972. Seasonal variation in the composition of the volatile oil of the leaves, buds, and twigs of white spruce (*Picea glauca*). Canadian Journal of Botany 50, 1595-1603.

Winer, A.M., Arey, J., Atkinson, R., Aschmann, S.M., Long, W.D., Morrison, C.L., Olszyk, D.M., 1992. Emission rates of organics from vegetation in California's central valley. Atmospheric Environment 26A (14), 2647-2659.

Yu, J., Cocker, D.R., III, Griffin, R.J., Flagan, R.C., Seifeld, J.H. Gas-phase ozone oxidation of Monoterpenes: gaseous and particulate products. J.Atm.Chem., 1999, 34,207-258.

Yu , J. *et al.*, Observation of Gaseous and Particulate of MOxidati Productsonoterpene on in Forest Atmospheres. *Geophys. Res. Lett.* 26 (1999).

Yokouchi, Y., Ambe, Y., 1985. Aerosols formed from the chemical reaction of monoterpenes and ozone. Atmospheric Environment 19 (8), 1271-1276

Zhang, J. Y., Hartz, K. E. H., Pandis, S. N., and Donahue, N. M.: Secondary organic aerosol formation from limonene ozonolysis: Homogeneous and heterogeneous influences as a function of NOx, J. Phys. Chem. A, 110, 11053–11063, 2006.

Zimmerman, P., 1979. Testing of hydrocarbon emissions from vegetation, leaf litter and aquatic surfaces, development of a methodology for compiling biogenic emission inventories. U.S. Environmental Protection Agency, NTIS No.PB296070, EPA-450/4-79-004, pp. 1-71.

Chapter 6

Separating Cloud Forming Nuclei from Interstitial Aerosol

Gourihar Kulkarni

Additional information is available at the end of the chapter

1. Introduction

Our poor representation of aerosol and cloud interactions in the climate models have led to the largest uncertainty in predicting climate change. Studies have shown that CN can influence climate by changing the properties of clouds. Aerosol particles that act as CN can be broadly classified based on their source into two categories: natural and anthropogenic aerosol. The global source strength of natural aerosol is higher than anthropogenic aerosol; however, certain specific anthropogenic constituents can amplify the aerosol effect on clouds. In addition, the atmospheric trace amounts of soluble gases and organic substances can alter the aerosol properties from both of the sources.

Recent studies have shown that various aerosol properties (size, surface chemistry [hygroscopicity and wettability] and active sites) as a function of temperature and humidity can determine the CN efficiency of aerosol. Atmospheric scientists are working towards finding a relationship between these properties to parameterize the observations in the climate models. But without an adequate knowledge of CN properties this representation cannot be improved further.

The technique of CN separation from the interstitial aerosol has the advantage that by measuring the specific properties of CN, simplifies the model representation task. For example, laboratory and *in-situ* techniques can be used to differentiate the CN in CCN and/or IN measurements and their properties can be measured. Therefore, the modelers can narrow down the physicochemical properties of CN to be incorporated into the representation task. Further, the information of the aerosol chemistry helps to determine aerosol source: natural versus anthropogenic.

2. Separation techniques

The separation of CN from interstitial aerosol technique is based on the particle's inertia. The instrument that employs this technique is called counterflow virtual impactor (CVI). The separation is achieved by stopping and removing the gas phase and small particles but capturing large particles with sufficient inertia to cross gas streamlines. Particles with insufficient inertia to be captured follow the deflected streamlines and are removed from the system. Higher inertia particles are injected into a typically clean, dry, and warm counterflow carrier gas that causes evaporation of condensed phase water. This technique has the advantage that a broad cut size range can be achieved by varying the flow rates associated with the CVI without changing the physical dimensions of the instrument.

The CVI used in the laboratory set up is called pumped CVI (PCVI) and the CVIs used for *in-situ* measurements are called airborne CVI (ACVI). The flow schematics of these designs are shown in Figure 1 a) and b), respectively. In PCVI design the aerosol particles are pulled inside the instrument and undergo inertial separation, while in ACVI the aircraft velocity imparts motion for aerosol particles that are again separated based on the inertia. Both designs are widely used and their performance characteristics are documented.

3. Design considerations

The CVI's performance is characterized by a particle collection efficiency curve. In an ideal environment, the separation between the CN and interstitial particles should be perfectly sharp. However, due to the non-idealistic flow behavior within the CVI, the true efficiency is hardly achieved.

Figure 2 shows the particle transmission efficiency (TE) of ammonium sulfate particles as a function of its size [1]. Three different flow configurations were used, implying the importance of relationship between the flows. It can be observed that by varying the flows, different sizes of particles can be sampled, but, as mentioned above, due to non-idealistic flow behavior, the TE do not reach 100%: imperfect TE and also sharp TE are not observed. It was suggested that imperfect TE is caused because the particles near the wall surface are trapped in the recirculation zone and do not join the sample flow. Also, because the flow within the counterflow region is not well-developed. This flow heterogeneity allows the small particles to penetrate the counterflow region and join the sample flow, even though they should be rejected.

The particle TE can be theoretically calculated based on the following equation [2],

$$L = K.r.\frac{\rho_p}{\rho_g}.\left(Re^{1/3}.C^{1/2} - \frac{\pi}{2} + \varphi\right) \tag{1}$$

Where,

$\varphi = tan^{-1}\left(Re^{-1/3}.C^{-1/2}\right); 0 \leq \varphi \leq \frac{\pi}{2}$, L is the stopping distance, K is a constant (= 5.3075), r is the radius of the particle, ρ_p is the density of the particle, ρ_g is the density of the flow media, Re is the Reynolds number, and C is a constant (= 0.158). For the desired flow configuration, if the distance between the tip of the CVI nozzle till the beginning of the sample flow is larger than the particle stopping distance, then the particle joins the sample flow and is transmitted. Figure 3 shows the relationship between the droplet diameters to the stopping distance for different flows. The flow can be varied either by increasing the input flow, while keeping other flows constant, or by varying the CVI geometry. The former option is always desired.

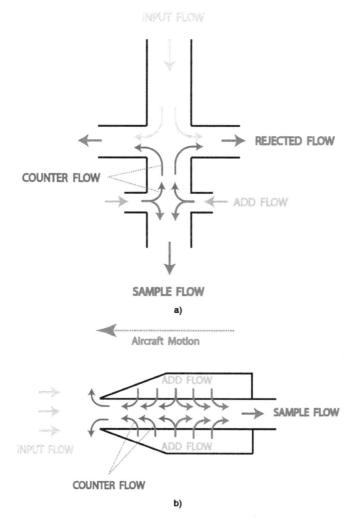

Figure 1. a: Flow schematics within the pumped counterflow virtual impactor (PCVI). Input flow carries the condensation nuclei and interstitial particles (non-activated aerosol particles). The large particles that have sufficient inertia to cross the streamlines enter the counterflow; these particles that

can overcome the counterflow, join the sample flow. The remaining particles join the rejected flow. The sampled condensation nuclei (CN) are then forwarded to the respective analytical tools. The add flow consists of dry and particulate-free gas that splits into counterflow and sample flow. The counterflow then joins the input flow to become the rejected flow. In normal PCVI operation, the sample flow is maintained constant while the counterflow is varied to sample various CN sizes. b: Flow schematics within the airborne counterflow virtual impactor (ACVI). The motion to the CN and interstitial particles is given by the motion of the aircraft. This induces the input flow and therefore, indirectly, the inertia to the particles. Similar to the PCVI operation, the counterflow rejects the particles that do not have sufficient inertia to overcome the counterflow; the particles having sufficient inertia join the sample flow. The sample particles are then forwarded to the desired analysis tools. Again, similar to the PCVI, the add flow splits into counterflow and sample flow, and thus, by varying the add flow, various sizes of CN can be sampled (assuming sample flow is maintained constant).

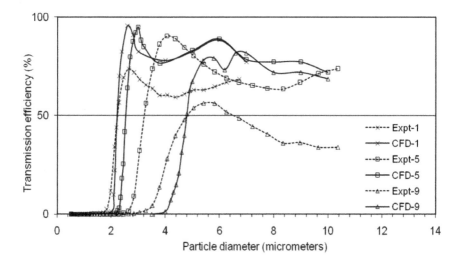

Figure 2. Particle transmission efficiency (TE) of ammonium sulfate particles. Experimental (Expt) TE are compared to computational fluid dynamics (CFD) predicted TE for various cases where PCVI flow configuration was varied [1].

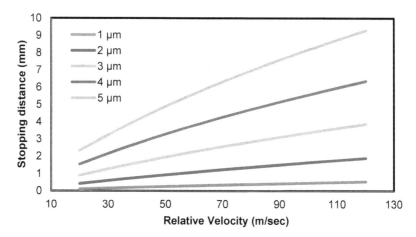

Figure 3. Stopping distance of various-size water droplets as a function of their relative velocity with respect to flow media when exiting the nozzle (in PCVI) or entering the counterflow region (in ACVI). If the physical dimensions of the CVI are known, then these relationships can be used to determine the theoretical TE of the particles.

4. Operational challenges

As discussed above, the PCVI performance characteristics depend upon flow behavior and geometrical design. Recently, limitations and uncertainties associated with the CVI's have been identified [1]. They include particle losses at walls, imperfect transmission efficiencies of CN, limited size range of transmitted particles, turbulence effect on the droplet breakup and shattering, and narrow range of measurement flow rates. To understand the artifacts and improve further the designs, CFD simulations were carried out. For example, fluid flow characteristics of PCVI are analyzed to understand the performance characteristics and associated artifacts, as shown in Figure 4.

The white colored particle deposition on the walls can be observed (Fig. 4 top panel). The particle deposition losses occur at various eddies and vortices shown in the bottom panel of Fig. 4. The magnified CFD image shows eddies and recirculation vortices generated as a result of the flow boundary conditions and the PCVI design geometry. Such CFD simulations are necessary to improve the design to reduce the particle losses.

Under the influence of flow turbulence within the CVI instrument, it is possible that when sampling cloud, hydrometeors (liquid droplet and ice crystals) can break or shatter leading to numerous small particles. This is undesirable as the particle TE will be reduced and might lead to non-conclusive results. It has also been observed that at high airspeeds (in airborne CVI), large drops and ice crystals can impact on the probe inlet and break; this also happens within the counterflow region (because of shock). Droplet breakup criterion is usually calculated using Weber number: $We = (V_g-V_{drop})^2 \cdot \varrho_g \cdot d_{drop}/\sigma$, where V is the velocity of the gas

(g) and droplet (d; drop). The droplet breaks when the We number is greater than 12. These estimates can be validated by combining the observations and CFD simulations.

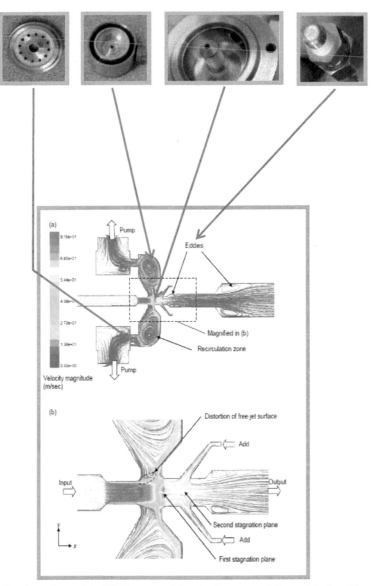

Figure 4. Interior surfaces of the PCVI instrument showing (upper panel) the particle deposition regions. These regions are indicated within CFD predicted velocity pathlines (bottom panel). See text for details [3].

Another feature that might be important, which is not well-documented, is the scavenging of the interstitial aerosols when sampling cloud droplets. Due to the mass differences between these two types of particles, the relative velocity could become significant and could lead to collision between the droplet and interstitial aerosol. If the droplet collides with the interstitial particles, then these particles might get trapped within the droplet and if this droplet gets transmitted, that will yield undesired results. As of now, non-activated aerosol particles (interstitial aerosol) are being characterized as activated aerosol particles, but this is not correct. Systematic studies are required where water droplets and interstitial aerosols should be generated and their collision efficiency should be investigated.

5. Summary

In this chapter, a technique that separates the cloud forming nuclei from the interstitial aerosols is briefly discussed. The technique is based on the inertia of the particle. Cloud forming nuclei are the residual particles of the droplets and ice crystals. These cloud hydrometeors have high inertia compared to the interstitial aerosols and therefore penetrate the counterflow region of the CVI to be sampled. Two types of CVI instruments are based on this technique: PCVI and ACVI. PCVI is generally used in the laboratory set-up where the particle velocity is achieved by pumping the input flow; whereas, in the ACVI, the particle velocity is generated by aircraft flight.

Transmission efficiency of the particles that are sampled can be theoretically calculated, and it was observed that as particle velocity and/or its diameter increases the efficiency also increases. Several artifacts of the cloud separation technique are described. They include particle losses and imperfect transmission efficiencies, flow turbulence effects on the droplet breakup and shattering, and possibility of scavenging of interstitial aerosols (this needs further investigation). However, many studies have quantified these artifacts and the cloud separation technique is now considered as a must have measurement platform for most of the laboratory and field studies.

Author details

Gourihar Kulkarni
Pacific Northwest National Laboratory, Richland, WA, USA

6. References

[1] Kulkarni, G., M. Pekour, A. Afchine, D. M. Murphy, and D. J. Cziczo: Comparison of experimental and numerical studies of the performance characteristics of a pumped counterflow virtual impactor, Aerosol Sci. Tech., 45:382–392, 2011

[2] Serafini, J. S.: Nat. Adv. Comm. Aeron., Report 1159, Impingement of Water Droplets on Wedges and Double-wedge Airfoils at Supersonic Speeds (1 960). NACA Report # 1

159, 40th Annual Report of the National Advisory, U.S. G.P.O. Washington, D.C. pp. 85-108, 1954

[3]   Kulkarni G. and Twohy, C.: Computational fluid dynamics studies to understand ice crystal and liquid droplet breakup within an airborne counterflow virtual impactor, AAAR 30th Annual Conference, 2010

Chapter 7

A Method Analyzing Aerosol Particle Shape and Scattering Based on Imaging

Shiyong Shao, Yinbo Huang and Ruizhong Rao

Additional information is available at the end of the chapter

1. Introduction

Aerosol particle shape is a key parameter affecting its physical characters, especially scattering properties[1]. The information of shape reveals important application in such fields as atmospheric radiation and remote sensing, climate research, radar meteorology[2]. The convenient availability and simplicity of the Lorenz-Mie theory has resulted in a widespread practice of treating non-spherical particles as if they were spheres to which Lorenz-Mie results are applicable. However, the assumption of sphere is rarely made after first having studied the effects of non-sphere and concluded that they are negligible but is usually based on a perceived lack of practical alternatives[3].

In a variety of occupational, environmental and industrial scenarios, particles within the size range from a few tenths of a micrometer to a few hundred micrometers play an important role[4]. Since the majority of aerosol particles are to some extent non-spherical and indicating relation with their origins, the knowledge of particles' shape may be used to judge the source of those particles and hence facilitate more effective contamination control and to reduce inadvertent particle generation. For example, fibrous particles are often corresponding to textile industry, flake-like particles corresponding to papermaking industry, etc.

The scattering profile of light scattered by any particle is determined by its size parameter, its shape, and its orientation with respect to the incident illumination[4]. The spatial intensity distribution of scattered light thus contains information by which the particle may often be classified or even identified. The light scattering suits to be used for deducing shape of aerosol particles by detecting scattering information, which is rapid and non-contact[5-7]. By analyzing pairs of signal from opposite detectors, Diehl differentiate bluffly the shape of suspending particles[8]. Bartholdi reflected majority of scattering light onto a circular photodiodes array, and gained more abundant information about particle shape[9]. Kaye assessed the feasibility of classifying individual aerosol particles on the basis of size and shape parameters, which determined by measurement and analysis of the spatial intensity distribution of laser radiation scattered by the particle shown in Fig.1[10,11].

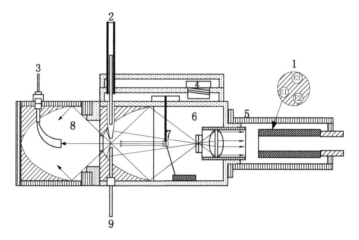

1.Detector Channels E1,E2,E3 2.Sample Inlet 3.Detector Channel E4 4.Filter for Sheath Air 5.Photomultiplier House 6.Main-Chamber 7.Laser and Modulating System 8.Rear-Chamber 9.Pump

Figure 1. Schematic diagram of the aerosol shape analyzer

The laser beam is directed onto the particle flow by a small 45° mirror supported by an optical window. Particle-laden air is drawn in through the scattering chamber in laminar flow and is ensheathed by filtered air drawn in through ports simultaneously[12]. Individual particles in the sample air transverse the laser beam and produce pulses of scattered light. Three miniature photomultipliers are incorporated an arrangement to allow measurement of variations in azimuthal scattering from individual airborne particles between 28 ° and 141 to the beam incident direction upon an ellipsoidal reflector whose primary focus is coincident with the scattering volume. The output of detector E4 measures the forwards scattering used in estimation of the particle size. The first developments to achieve this are incorporated in Biral's ASAS™ technology and has been developed by the UK armed forces. The instrument allows airborne particles in the sub-10 μm size range to be classified into size and shape classes in real-time at rates of up to about 10000 particles per second.

Although the instruments described above are able to differentiate between spherical and non-spherical particles, also provide some crude indication of particle shape, the full potential of spatial intensity scattering analysis for non-spherical particle characterization should only be realized by the detailed analysis both of azimuthal and of polar scattered intensity variations.

To observe micro-particle, the microscopy is often preferred instrument, which represents an excellent technique for directly examining target[13]. However, for manual microscopy, elaborate sample preparation is necessary and only a few particles can be examined resulting in very low statistical relevance of the data. Recently, a faster evaluation of activated sludge floc properties became possible by connecting the microscope to automated image analysis software[14], see Fig.2.

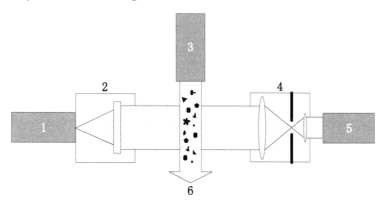

1. Light Source 2.Light Modulation 3.Sample Dispersion 4.Imaging Object Lens 5.CCD with microscope 6.Particle Stream

Figure 2. Sketch map of the microscope CCD system

The light of a pulsed light source is expanded by a beam expansion unit, which creates a parallel beam. The dispersed particle flow is illuminated and finally imaged by an CCD via microscope. The particles show arbitrary orientation and the number of overlapping particles are lost. The light source creates stable visible light pulses about 1 ns at power of about 0.15 nJ/pulse. The repetition rate is adjustable from 1 to 500 Hz meeting the specifications of the high speed CMOS camera. One image is composed out of 1024×1024 pixels of $10 \mu m \times 10 \mu m$ area with 256 gray levels. Imaging objectives for different magnifications are mounted on a carousel for simple selection of a measuring range by software.

Light scattering and imaging by CCD via microscope are routine two methods for detecting aerosol particle shape. CCD video microscope is volume-based, and light scattering is number -based[15- 17]. If a number distribution fits a log-normal distribution, then its transformation to volume distribution will result in another log-normal distribution, characterized by[18]:

$$\ln x_g \bullet V = \ln x_g \bullet N + 3\ln^2 \sigma_g \tag{1}$$

x_g is the geometric mean of the distribution, σ_g is the geometric standard deviation, V is volume based distribution and N is number based distribution.

The angular scattered light intensity largely depends on the optical properties of the particles. For small particles whose radiuses lower than $10 \mu m$, the refractive index dependence becomes significant because at such small sizes the light irradiated onto the particle is not completely absorbed and can emerge as a refracted ray. In this case, the Mie theory should be used instead of the Fraunhofer theory, which does not take into account the optical properties of the particles. When examining the activated sludge floc size, the optical poly-disperse properties are difficult to be characterized and the Fraunhofer theory has to be used. The section from Mie to Fraunhofer needs to be revised by other method, beyond all question, the image technique is good approach.

The techniques based on laser light scattering are more suited to follow the fast changes that may occur in floc size during the process. Since the light scattering method doesn't usually offer visual information, coupling it to an image analysis system allows a direct visual inspection of the process evolution. If combine light scattering and CCD video microscope, not only the classification of particle shape can be realized, also the comparison and analysis of results between experiment and calculation by corresponding shape can bring more information which impossible received by individual method. The chapter describes a new instrument, aerosol particle shape and scattering analyzer based on imaging. By analyzing scattering intensity coefficient and polarization of fibre cotton and calculation from wave theory, the affecting factors are pointed out.

2. Description of the instrument

Figure 3 shows the experimental apparatus to measure the shape and scattering properties of aerosol particles in analog manner at the semiconductor laser wavelength of 0.65um. The instrument realizes the combination of imaging and light scattering[19]. The scattering chamber, a homocentric hollow black sphere showed in Fig.3, is the core portion of the analyzer based on imaging. The hollow black sphere is composed by two symmetrical hollow hemispheres, which fabricated of aluminium considering hardness and weight. The interior wall of the hollow sphere is made coarse elaborately to reduce the influence from its scattering and reflection. The chamber inner diameter is 48mm, and the outer diameter is 76mm. There is a large aperture at top and bottom on the vertical radial line of the chamber, aerosol particle inlet and aerosol particle outlet respectively. The apertures at front and back on the horizontal radial line are respectively for assembling semiconductor laser and CCD video microscope. Without saturation of CCD, there is a filter corresponding laser wavelength in front of microscope. 36 small apertures for optical fibre that are positioned to measure the light from the horizontal and vertical scattering angles between 30° and 150° in 15° increments. The diameter of each aperture is 3.02mm, which is slightly greater than diameter of optical fibre. The laser plane of polarization is set perpendicular to the horizontal plane. The inner diameter of the aperture port near the interior wall of scattering chamber, whose thickness is 1mm, is a little smaller than outer diameter of optical fibre, and the all ports for collecting scattering light can be strictly same distance from the centre of the chamber. The axeses of all apertures are directing the centre of the scattering chamber. The respective horizontal and vertical 18 small apertures are symmetrical about horizontal radial line. Each optical fibre collects the scattering light with an acceptance angle of $\pm 1.5^o$ since assembled a convex lens with 2.5mm diameter. Unfortunately, the small apertures on the vertical radial line happen to be concurrent with sample pipe and waste pipe.

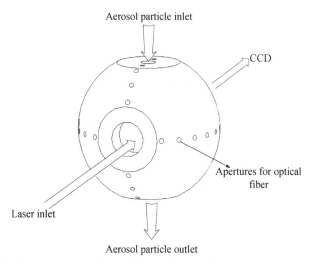

Figure 3. Simplified schematic map of aerosol particle shape and scattering analyzer

The scattering chamber must be puffed by clean air to eliminate the influence from the impurity. Single aerosol particle stream vertical to the laser beam is drawn in through the scattering chamber along the axes of aerosol particle inlet and outlet, and is ensheathed by filtered air drawn in through ports simultaneously. A set of filters and regulators introduce aerosol particles entrained in a fine laminar stream through the center of the chamber and intersecting the laser beam one particle at a time. Individual particles in the stream produce pulses of scattered light, which are amplified by photomultiplier tube detectors connected to a corresponding optical fibre bundles.

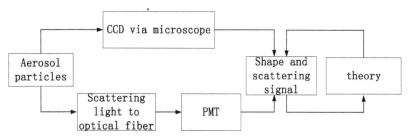

Figure 4. Working process of aerosol particle shape and scattering analyzer

The whole working process is described as Fig 4. The integration time of CCD is slightly less than 0.005s. During experiment, the laser beam diameter is 1.5mm, and wavelength is 650nm. When single particle stream passing chamber center, the image is immediately acquired by the CCD video microscope, at the same time, the scattering light of corresponding particle is collected by optical fibers and transmitted to PMT. The speeds of aerosol particles are controlled by pressure difference of inner and outer hollow sphere. By adjusting the pressure difference, the particle speeds can be restrained less than $0.4\,m/s$ for effective diameter higher than $1\,\mu m$.

Since scattering light contains the information about shape of particle, more significant conclusion can be obtained by comparing experimental results and calculation from theory. The shapes of aerosol particles can be deduced through scattering light distribution, and the result will be Verified by corresponding images from CCD. So the data library of scattering and image about aerosol particles is gradually built, moreover, the aerosol particles are classified according to relevant shape and size.

3. Result and discussion

When the aspect radio exceeds 10, fiber particle, a common shape sort in aerosol, can be considered infinitely long cylinder. Generally, particles with effective radius less than 10um are inhaled[20]. However, in 1990, the U.S. National Institute for Occupational Safety and Health stated "no evidence for a threshold or 'save' level of asbestos exposure[21]. Flying particles are inclined to keep their long axes consistent with axes of carried gas, which ensures scattering light relatively steady.

3.1. Wave theory for infinitely long cylinder

The scattering geometry of infinitely long cylinder is shown in Fig.5. The z axis of the cylindrical coordinates (r,ϕ,z) is defined along the central axis of the cylinder. The angle between the incident ray and the negative z axis is denoted as χ. α is defined as an oblique incident angle which is the complement angle of χ. The x axis is defined in the plane containing the direction of the incident ray and the z axis. This plane defines the angles $\phi=0$ and $\phi=\pi$. The coordinate r is then contained on the xy plane such that the cylinder occupies the region $r\leq a$, where a is the cross-section radius of the cylinder[22,23]. To illustrate the scattering geometry, a cylinder whose diameter is larger than the incident wavelength λ so that the geometric optics will be used. The rays externally reflected, refracted, and internally reflected on the surface of the cylinder follow the Snell laws.

The scattering angle θ, which is defined as the angle between the direction of the incident wave and the scattered wave, is obtained:

$$\cos\theta = \sin^2\alpha + \cos^2\alpha\cos\phi \qquad (2)$$

ϕ is defined as an observation angle to distinguish it from the scattering angle. ϕ and θ are equal only at normal incidence. In all other cases, the values of ϕ are always more than that of θ. So there is no true backscattering for an infinitely long cylinder. Starting from

Maxwell equations, after complex algebraic operations, the scattering coefficients a_n and b_n are deduced as below:

$$b_{n1} = P_n \frac{Q_n^2 + A_n(\varepsilon_1)B_n(\varepsilon_2)}{Q_n^2 + A_n(\varepsilon_1)A_n(\varepsilon_2)}$$

$$a_{n2} = P_n \frac{Q_n^2 + B_n(\varepsilon_1)A_n(\varepsilon_2)}{Q_n^2 + A_n(\varepsilon_1)A_n(\varepsilon_2)} \qquad (3)$$

$$a_{n1} = -b_{n2} = P_n Q_n \frac{A_n(\varepsilon_1) - B_n(\varepsilon_1)}{Q_n^2 + A_n(\varepsilon_1)A_n(\varepsilon_2)}$$

Where

$$A_n(\varepsilon_{1,2}) = j \frac{H_n^{(2)'}(la)}{H_n^{(2)}(la)} - \varepsilon_{1,2} l \frac{J_n'(ja)}{J_n(ja)}$$

$$B_n(\varepsilon_{1,2}) = j \frac{J_n^{(2)'}(la)}{J_n^{(2)}(la)} - \varepsilon_{1,2} l \frac{J_n'(ja)}{J_n(ja)}, \begin{cases} \varepsilon_1 = 1 \\ \varepsilon_2 = m^2 \end{cases}$$

$$P_n = J_n(la) / H_n^{(2)}(la) \qquad (4)$$

$$Q_n = inh(l^2 - j^2) / xlj$$

$$x = ka = 2\pi a / \lambda$$

If $\alpha = 0^\circ$, then $a_{n1} = b_{n2} = 0$. It should be noted that these coefficients depend on the refractive index, the size parameter and the oblique incident angle.

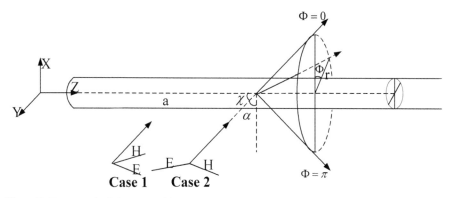

Figure 5. Geometry for light scattered by an infinitely long cylinder

Now we shall consider two simple cases separately. First, the electric vector $\vec{E}$ is parallel to the incident plane. This is sometimes called the TM mode. For the second case, the electric vector $\vec{E}$ is perpendicular to the incident plane and called the TE mode. The intensities of the scattered light in any direction are:

$$I_{TM-11} = 2i_{11}I_0 / \pi kR$$
$$I_{TM-12} = 2i_{12}I_0 / \pi kR$$

$$I_{TE-22} = 2i_{22}I_0 / \pi kR$$
$$I_{TE-21} = 2i_{21}I_0 / \pi kR \tag{5}$$

The intensity coefficients for above two cases are defined as:

$$\text{TM}\begin{cases} i_{11} = \left| b_{01} + 2\sum_{n=1}^{\infty} b_{n1} \cos n\phi \right|^2 \\ i_{12} = \left| 2\sum_{n=1}^{\infty} a_{n1} \sin n\phi \right|^2 \end{cases} \qquad \text{TE}\begin{cases} i_{22} = \left| a_{02} + 2\sum_{n=1}^{\infty} a_{n2} \cos n\phi \right|^2 \\ i_{21} = \left| 2\sum_{n=1}^{\infty} b_{n2} \sin n\phi \right|^2 \end{cases} \tag{6}$$

where a_{n1}, b_{n2}, a_{02} and b_{01} are scattering coefficients. i_{11} and i_{22} are the scattered intensities that lies in the same plane as the incident intensities, while i_{12} and i_{21} are the cross-polarized scattered intensities that have directions perpendicular to the incident intensities, what's more, $i_{12} = i_{21}$.

The polarization of scattering light is defined as[24]:

$$\text{TM:} \quad p_{11} = \frac{i_{11} - i_{12}}{i_{11} + i_{12}} \qquad \text{TE:} \quad p_{22} = \frac{i_{22} - i_{21}}{i_{22} + i_{21}} \tag{7}$$

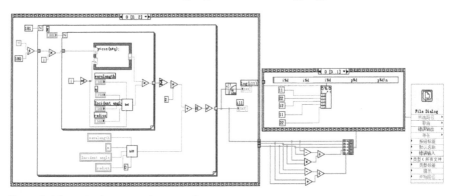

Figure 6. Schematic of programming wave theory

LabVIEW is a graphical programming language which has its roots in data acquisition and automation control. Its graphical representation, similar to a process flow diagram was created to provide an intuitive programming environment for users[25]. The language has matured over the last twenty years to become a general purpose programming environment. With LabVIEW, we have self-programmed wave theory as a part of the whole measuring system. The programming structure is demonstrated partly in Fig.6. The program consists of 3 relevant parts, which are similar at format. The calculation data, which is function of incident angle, refractive index, incident wavelength and cylinder diameter, is saved in form of .txt. The data will be more analysed through Origin software and compared with information from structure according to shape given by CCD via microscope.

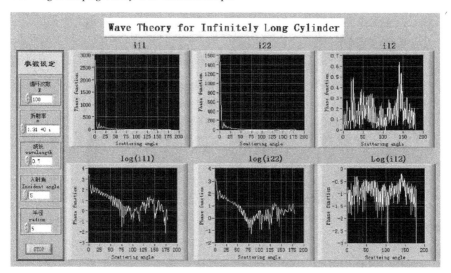

Figure 7. Interface of programming wave theory

The interface of the program for calculation is shown in Fig.7, which contains input parameter area and display windows of calculating results. The display windows contain i_{11}, i_{22} and i_{12}, also their logarithmic format. The input parameters include refractive index, laser wavelength, incident angle and radius of cylinder. In a general way, the whole process for calculation is shorter than 30s. With the increase of incident angle, the calculation time will be extended slightly.

3.2. Analysis of experiment and calculation

The effective radius of selected fiber cotton particles for experiment are about 10um, compared to 1mm laser beam, the condition of infinitely long for irradiated cotton is satisfied. The incident laser with 0.65um wavelength is transformed to linear polarized light by Glan-Talyor lens. When electric vector $\vec{E}$ is parallel to the incident plane, the refractive index of fibre cotton is 1.573-1.581, we choose the middle number 1.577 for calculation. When the electric vector $\vec{E}$ perpendicular to the incident plane, the corresponding refractive index is 1.524-1.534, we also choose the middle number 1.529 for calculation. In experiment, the angle between cotton and axes of carried gas is about 5°.

The "left, right, up and down" in figure 4 refer to figure 1.It can be concluded that the tendency of experiment data keeps uniform with calculation of the infinitely long cylinder for scattering intensity and polarization. A pair of experimental signal lack in polarization

P_{11} and P_{22}, since up aperture and aerosol particle inlet with a same positions, similarly to bottom aperture and aerosol particle outlet, too. i_{11} and P_{11} are more close to calculation than i_{22} and P_{22}, which might be caused by different outline of fibre cotton particles. Clearly, the difference between cotton fibre in image A and infinitely long cylinder is smaller than that between image B and infinitely long cylinder. The experiment is cursory in describing trendy of scattering intensity and polarization restricted by the number of optical fibre for collecting scattering light; on the one hand, the angle increments between apertures for optical fibre are limited in manufacturing process; on the other hand, the cone angle of receiving plane for every optical fibre is 3°, unlike the elements of calculation, the integral photometric characteristics are much less dependent on particle shape.

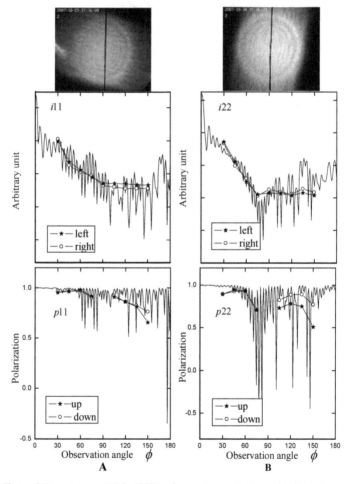

Figure 8. Photo of fiber micro-particle by CCD and respective scattering and polarization

4. Conclusion

An experimental apparatus has been built to measure the images and light scattering characteristics of aerosol particles simultaneously. The core portion of the analyzer is a homocentric hollow black chamber. Images, corresponding scattering intensity and polarization of fiber cottons are received. Wave theory for infinitely long cylinder has been compiled with LabVIEW. By comparison of experimental data and calculation, the affecting factors to results are pointed out, which provides a good foundation to further study.

Author details

Shiyong Shao*, Yinbo Huang and Ruizhong Rao
Key Laboratory of Atmospheric Composition and Optical Radiation, Anhui Institute of Optics and Fine Mechanics, Chinese Academy of Sciences, China

Acknowledgement

The authors are very thankful to the reviewers for valuable comments. This work was supported by Youth Talent fund from Hefei Institutes of Physics Science under Contract No. Y03AG31141. The theoretical calculations in this paper have been kindly assisted by Lei Hao and Yongbang Yao helped with the construction of the apparatus.

5. References

[1] Shao, S.Y., Huang, Y.b., Yao, Y.b., Rao, R.Z., "Progress in Shape Measurement Technology of Micro-particles in Atmosphere, " Journal of Atmospheric and Environmental optics 3(1), 1-10(2008).

[2] Mishchenko, M.I., Lacis, A.A., "Scattering, absorption, and emission of light by small particles, " Cambridge university press, New York, 279-359 (2002).

[3] Zhao, J.qi, Shi, G.Y.., "Light Scattering Properties of Snall Nonspherical Particles, " Science Technology and Engineering., 5(24):1872-1875 (2005).

[4] Edwin, H., Paul, H.K., John, R.G.., "Light scattering from non-spherical airborne particles: experimental and theoretical comparisons, " Appl. Opti., 33(3), 7180-7186(1994).

[5] Mi, F.W., "The laser diffraction particle size analyzer for methods to calculate particle size distribution, " Acta Photonica Sinica, 28(2), 151-154(1999).

[6] Li, L.F., Zhang, L., Dong, L., "Experimental Study of the Concent ration of Soot Based on the Method of Optical Back Scattering, " Acta Photonica Sinica, 35 (6), 9152918(2006).

[7] Liu, J., Hua, D.X., L I, Y., "Ultraviolet lidar for profiling of t he urban atmospheric aerosol spatial and temporal at Xi'an, " Acta Photonica Sinica, 36 (8), 153421537(2007).

[8] Diehl, S.R., Smith, D.T., Sydor, M., "Analysis of suspended solids by single-particle scattering, " Appl.Opt, 18(10), 1653-1658 (1979).

---

* Corresponding Author

[9] Bartholdi, M., Salzman, G.C., Hiebert, R.D., Kerker, M., "Differential light scattering photometer for rapid analysis of single particles in flow, " Appl.Opt, 19(10), 1573-1581(1980).

[10] Kaye, P.H., Eyles, N.A., Ludlow, I.K., "An instrument for the classification of airborne particles on the basis of size, shape and count frequency, " Atmospheric environment, 25(3), 645-654(1991).

[11] Kaye, P.H., "Spatial light-scattering analysis as a mean of characterizing and classifying non-spherical particles, " Meas.Sci.Technol, 9, 141-149(1998).

[12] Foot, V.E., Clark, J.M., Baxter, K.L., Close, N., "Characterising single airborne particles by fluorescence emission and spatial analysis of elastic scattered light, " in Optically Based Biological and Chemical Sensing for Defence. Proc.SPIE., 5617, 222-299(2004).

[13] Barbusinski, K., Koscielniak, H., "Influence of substrate loading intensity on floc size in the activated sludge process, " Wat.Res., 29(7), 1703-1710(1995).

[14] Grijspeerdt, K., Verstraete, W., "Image analysis to estimate the settleability and concentration of activated sludge, " Wat.Res., 31(6), 1126-1134(1997).

[15] Arjen Van Der Schoot, "Dual-channel particle size and shape analyzer, " China Particuology, 2(1), 44-45 (2004).

[16] Govoreanu, R, Vandegehuchte, K, Saveyn, H, "An automated image analysis system for on-lines structural characterization of the activated sludge flocs, "Med. Fac. Landbouww. Univ. Gent" 67(4), 175-178(2002).

[17] Govoreanu, R., Saveyn, H., Meeren, P.V.D, Vanrolleghem, P.A., "Simultaneous determination of activated sludge floc size distribution by different techniques, "Water Sci. Technol., 50(12), 39–46(2004).

[18] Allen T., "Particle size measurements, "Fifth edition. Chapman and Hall Ltd., London.

[19] Shao, S.Y., Yao, Y.b., Rao, R.Z., "New instrument for detecting shape and scattering of micro-particles based on imaging, " China Patent, No:200710023960.X.

[20] Liu, J.J., Zhang, W.G., "Design and Investigation of Method for Accuracy Determination of Inhalable Particulate Matter Sampler, "China powder science and technology, 5(12):5-8(2006).

[21] Paul, K., Edwin, H., Zhenni, W.T., "Neural-network-based spatial light-scattering instrument for hazardous airborne fiber detection, " Appl. opt, 36(24), 6149-6156(1997).

[22] Liou, K.N., "Electromagnetic scattering by arbitrarily oriented ice cylinders, " Appl. Opti., 11(3), 667-674 (1972).

[23] Van de Hulst H.C., "Light scattering by small particles, " John Willey & Sons, New York, 297-328(1957).

[24] Zhao, K.H., Zhong, X.H., "Optics"(Volume one). Beijing: Beijing University Press, 235-245(2004).

[25] Zhou, Y.H., Wang, Y.N., Wu, W.C., "Design and implementation of voltage-current testing system based on LabVIEW for photovoltaic cells, " Chinese Journal of Scientific Instrument, 27(6), 1775-1776(2006).

Chapter 8

Experiences with Anthropogenic Aerosol Spread in the Environment

Karel Klouda, Stanislav Brádka and Petr Otáhal

Additional information is available at the end of the chapter

1. Introduction

The general public is well aware of harmful effects of solid contaminants in the atmosphere. The harmful effects depend on both the size and composition and origin of the particles. Solid particles greater than 100 micrometers remain in the air only very shortly and settle as dust. Smaller particles remain in the air substantially longer and may be transported through space. Particles smaller than 5 micrometers demonstrate aerosol properties and remain suspended in the air.

The inhalation of aerosols made of micro- and nanoparticles results in their deposition in the human respiratory system. It is expected that, depending on their diameter, surface, chemical composition of the surface etc., they are subsequently transported to other terminal organs.

There are many epidemiological studies that have identified the negative effects of these particles on respiratory and cardiovascular systems in sensitive members of the population [1]. A particularly serious effect on the cardiovascular system has been identified for inhaled ultrafine (nano) particles.

The main sources of dust, micro and nanoparticles that exceed natural background levels are anthropogenic activities, e.g. heavy industry, operations that involve metalworking and woodworking, milling, grinding, general dusty operations, etc. [2]. This has also been one of the main reasons for our measurements. The first part of the chapter presents the results of pilot and orientation measurements of means of transport in Prague, of an office building in the center of Prague, the influence of a Diesel engine type on the quantity of nanoparticles released into the atmosphere, and particles released during fire, welding, the burning of entertainment pyrotechnics, and the shooting of police weapons.

The second part of the chapter presents the results of systematic and long-term measurements of the quantities and distribution of nanoparticles at the platform of the busiest subway station in Prague, in a cabinet-maker workshop during processing of exotic woods, and in steelworks processing raw iron using the converter method.

Nonetheless, the results of the above-mentioned experiments should be viewed as results obtained at a particular time and place. The conclusions may be associated only with the specific situation. It is nearly impossible to obtain reproducible results, which is one of the main obstacles to the standardization of nanoparticle quantity in connection with their impact on human health (toxicity) and the environment.

2. Part I - Results of quantity and distribution measurements of aerosol nanoparticles at selected anthropogenic sources

2.1. The course of the experiments

2.1.1. Type experiment I – Prague subway

The measurement of nanoparticles was conducted inside a Prague subway train travelling on the C line during its regular operation with passengers and from the terminal station Letňany to the terminal station Háje and back. The measuring technology was situated in the 2nd (or the 4th) car of the train, on a seat in the outer part of the car.

2.1.2. Type experiment II – City bus

The measurement of nanoparticles was conducted in Prague in a city bus on line No. 189, travelling from the terminal station Sídliště Lhotka to the terminal station Kačerov, and after a break the bus travelled back to the station Sídliště Lhotka. The traffic level was 2-3 (i.e. partly traffic jams). The bus model was a Karosa B941 with a Liaz ML 636 engine. In the course of the measurement the occupancy rate of the bus fluctuated and reached a maximum of 60% of the bus capacity. The measuring technology was situated on a back seat.

2.1.3. Type experiment III – Car

The measurement of nanoparticles was conducted in a car that travelled essentially the same route as in experiment II and used various ventilation regimes. The car was a Skoda Octavia 1.6 with a gasoline engine, and the measuring technology was situated at the back of the car on the floor; there were 3 people travelling in the car.

2.1.4. Type experiment IV – Office building

In this case the measurement of nanoparticles was conducted in an office building in the center of Prague, situated at the corner of Dlážděná Street and the Senovážné Náměstí square. The measurements were conducted in several rooms in various locations on the building's layout, at various vertical levels, and for various types of operations.

2.1.5. Type experiment V – Simulated fires

The measurements of nanoparticles were conducted at simulated fires with various compositions of burning components in an open area.

Compositions of the burning pile were:

a. 3 straw mattresses, feather blanket, bed sheets, electric cable ca. 2 m, polystyrene ca. 1 m^2, dry wood from pruning natural seeding greenery;

b. 2 tires, polystyrene ca. 1 m^2, rubber hoses ca. 2 m, spent engine oil 10 l, Diesel oil 5 l, penetration paint 5 l, wood edgings.

The measuring technology for experiment V was situated 5 m from the fire edge. The aerosol samples were taken 0.5 m above ground level.

2.1.6. Type experiment VI – Diesel engines

Measurements of nanoparticles were performed for a Diesel engine and for a modern, environment-friendly Diesel engine.

a. The measured engine type was a Z 7701 Zetor Brno, 1600 rev., stroke volume 3922 cm^3, used in old tractor technology, mining engines, etc. The engine was put into operation in a testing room for combustion engines in DIMO Kamenná (Figure 1) and the measurements were performed at the outlet in front of the building. The distance of the measuring device from the outlet was 3 m.

b. The measured engine type was a part of a FORD –TRANSIT type FDG6 with the engine type PGFA, stroke volume 2198 cm^3, year of manufacture 2009. The measurements were performed with the engine running in neutral gear. The measuring device was situated 3 m and subsequently 7 m from the exhaust pipe.

2.1.7. Type experiment VII – Entertainment pyrotechnics

The measurements of nanoparticles were conducted at a simulated fireworks event that used various entertainment pyrotechnics available (mega cracker, fire hornet, sparklers, Bengal light, mega California, fire fountains, etc.) on a free area (street, square, etc.). The measuring technology was situated 12 m from the area where the entertainment pyrotechnics was gradually ignited.

2.1.8. Type experiment VIII – Welding in a workshop

The measurements of nanoparticles were conducted in a non-ventilated maintenance workshop (ca. 70 m^3). The welded product was a steel T-section 25 x 350 mm, welded with electrodes E-B 121, E 7018, SF 026126. The measuring device was situated 2.5 m from the welding location.

After the welding was completed (ca. 5 min.) the workshop was left without any activities, then the coagulation and sedimentation of particles was measured.

2.1.9. Type experiment IX – Shooting products

The measurements of nanoparticles in shooting products were conducted for weapons used by the Czech Republic Police (handgun CZ 75 D COMPACT, machine gun H&K MP5 KA4, shotgun Beneli M2, revolver King Cobra) at an open shooting range under real conditions. The gun muzzle was situated 0.5 – 0.7 m from the measuring device; the sample collection point and the gun muzzle were situated at the same height (Figure 2).

Figure 1. Tested Diesel engine type ZETOR

Figure 2. Measurement of handgun shooting

2.2. Results of the measurements and discussion

We have formulated the following conclusions from the obtained results:

- It is not possible to positively define a small increase in the quantity of nanoparticles depending on the occupancy rate of a subway car. This may be also affected by the surface locations of the stations and the subway route which may be situated close to an arterial road with busy traffic (ventilation shafts); see Figure 3 A) and 3 B).
- Passengers on the city bus line 189 were exposed to concentrations of nanoparticles higher by one order of magnitude than passengers on the subway line C (max. $36.7.10^3$ N/cm^3 in the subway, 260.10^3 N/cm^3 on the bus). However, even in this case it is impossible to prove a positive influence of the number of passengers on the concentration of nanoparticles; the effect of traffic density and traffic composition has been considered a more likely factor.
- The effect of traffic density has been demonstrated by measurements in the car, where the quantity of particles increased in the proximity of a slip road to the Prague ring and the highway D1 (location Kačerov).
- A certain protection of persons against nanoparticles in a driving car may be ensured by the reduction of ventilation and a pollen dust filter.
- Probably the most risky particles are those with a size up to 50 nm, as they are able to penetrate a protective cell barrier. Particles of that size were primarily identified in the proximity of the Kačerov location, which is again in the proximity of the slip road to the Prague ring and the highway D1. In the subway, the most exposed section (although the levels were much lower than on the bus) was in the center of Prague near the main arterial road (stations Pankrác – Florenc).
- Quite alarming was the finding that particles smaller than 40 nm were the most abundantly ascertained by the measurement on the city bus.
- The number of nanoparticles in the working premises of the office building occupied by non-smokers and used for standard office activities was slightly lower than in the building surroundings (ca. units $.10^3 N/cm^3$).
- Measurements performed in working premises on the individual floors in the building wing in the Dlážděná indicated no relationship between the quantity of nanoparticles and the floor altitude.
- An extreme increase of the number of nanoparticles was found in working premises where smoking was regular. For 3 people smoking at the same time, the number of nanoparticles increased by up to two orders of magnitude (although the levels were comparable with those on the city bus).
- Increase of the number of nanoparticles has been also demonstrated in a regular maintenance workshop.
- During fires, and during their extinguishing, the number of aerosol particles with nano dimensions strongly increased depending on the composition of the burning items (the measuring device was overloaded with fire products of mostly oil origin). The chemical composition of nanoparticles formed as burning products can be only speculated on [2].
- The increase of the overall quantity of aerosol nanoparticles for the classic Diesel engine was higher than that for the modern one, but the modern Diesel engine demonstrated an increase of nanoparticles in the dimensions that represent higher risks for human health and the environment (for comparison see Figures 4 and 5).

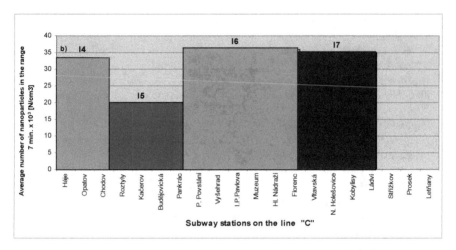

Figure 3. A) Graphic rendering of the number of nanoparticles in a subway car

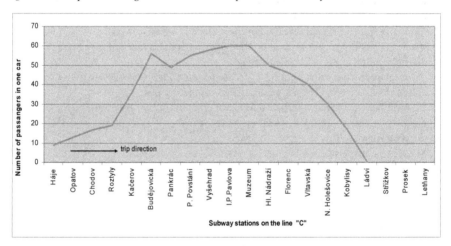

Figure 3. B) Graphic rendering of the depending on the location and number of passengers

- We have demonstrated that the concentration and size of nanoparticles change depending on the distance from their sources (see Figure 5). This is caused by dispersion and, particularly, by the coagulation of the particles (aggregation, agglomeration and adsorption of nanoparticles on microparticles, etc.).
- High-risk nanoparticles, in terms of size, have also been found for the classic engine before the engine was heated to the operating temperature (see Figure 4).
- Firefighters in action are threatened not only by particles generated by the fire and its extinguishing, but also by nanoparticles potentially generated by the firefighting technology, specifically the vehicles, Diesel aggregates, etc. (see Figure 6).

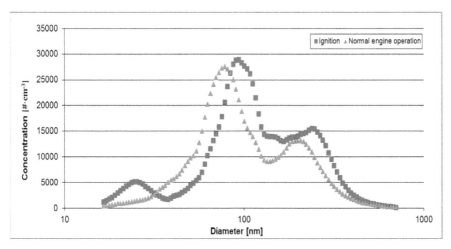

Figure 4. Distribution of particles measured at the Diesel engine made by ZETOR, exp. VI.a)

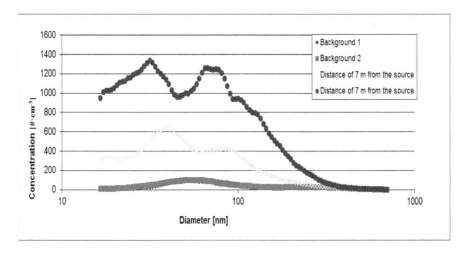

Figure 5. Distribution of aerosol particles measured at the exhaust pipe FORD TRANSIT exp. VI.b)

- The increase of nanoparticles after the ignition of entertainment pyrotechnics is two orders of magnitude higher than the background levels (see Figure 7, Table 1).
- The size of the nanoparticles from burning entertainment pyrotechnics was greater than 100 nm, while the dimensions were measured relatively far from the source. The time dependence of coagulation was also visible here; see the comparison of spectrums 2 and 3 in Figure 7.

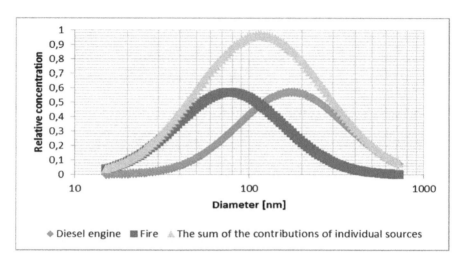

Figure 6. Distribution of particles exp. V.a

- During shooting, the overall concentration of aerosol particles per cm² increased, up to 200x in comparison with the initial background before shooting. Shooting products were released into the environment mainly from the weapon, the cartridge chamber, and ejected cartridge cases, and they spread in the shooting direction, to the sides and backwards.
- The highest concentration of aerosol particles after the shootings, more than double in comparison with the other weapons, was measured for the King Cobra revolver, while the spectrum had a pulsating character – see Figure 10 – and we have explained the high level of concentration of nanoparticles in after-shooting products also by the release of the products between the bullet and the muzzle.
- Dangerous nanoparticles smaller than 100 nm were generated immediately during welding, and they subsequently coagulated (see Figure 8).
- The speed of coagulation and sedimentation of particles during welding is obvious from the curve (see Figure 9, Table 2). It took essentially 3 hours before the background in the workshop dropped to the level before the welding.

The presented results of the performed experiments have only the character of basic measurements. It is very difficult to measure the number of nanoparticles, and the results are influenced by many factors (e.g. air flow, temperature, humidity, distance from the source etc.), as explained in the introductory section. It is practically impossible to get reproducible results of measurements, and this is one of the main problems of standardization of nanoparticles in respect to their impact on human health (toxicity) and the environment [3]. For these reasons, the conclusions provided herein may be associated only with the specific situations.

	Spectrum identification	Concentration of particles/cm³	Total weight of particles	Total volume of particles	Total surface of particles
			$\mu g/m^3$	nm^3/cm^3	nm^2/cm^3
Background	1	1530	6.44	5.37×10^9	1.18×10^8
Application of pyrotechnics	2	212000	1.27×10^3	1.06×10^{12}	2.77×10^{10}
Application of pyrotechnics	3	124000	1.79×10^3	1.49×10^{12}	2.56×10^{10}
No pyrotechnics	4	3290	15.5	1.3×10^{10}	2.77×10^8

Table 1. The physical values of nanoparticles measured during entertainment pyrotechnics experiments

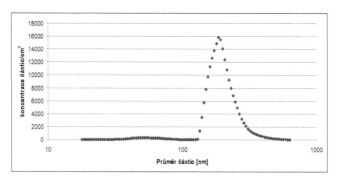

Spectrum No. 2 (spectrum identification see Table 1)

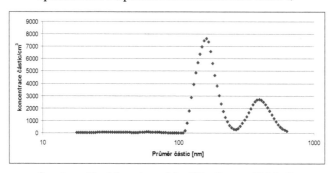

Spectrum No. 3 (spectrum identification see Table 1)

Figure 7. Spectra describing distribution of nanoparticles during entertainment pyrotechnics experiments. (Axis y: concentration of particles/cm³; axis x: diameter of particles [nm])

	Spectrum identification	Concentration of particles/cm³	Total weight of particles µg/m³	Total volume of particles nm³/cm³	Total surface of particles nm²/cm³
background	1	4620	14	1.17×10^{10}	2.9×10^{8}
welding	2	256000	7240	6.03×10^{12}	8.4×10^{10}
coagulation	3	333000	9910	8.26×10^{12}	1.21×10^{11}
coagulation	4	166000	5580	4.65×10^{12}	6.65×10^{10}
coagulation	5	98800	3570	2.98×10^{12}	4.18×10^{10}
coagulation	6	68900	2680	2.24×10^{12}	3.08×10^{10}

Table 2. The physical values of nanoparticles measured during welding experiments

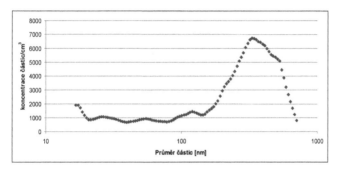

Spectrum No. 2 (spectrum identification see Table 2)

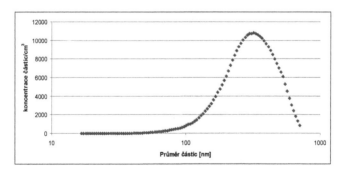

Spectrum No. 3 (spectrum identification see Table 2)

Figure 8. Spectra describing distribution of nanoparticles during welding experiments. (Axis y: concentration of particles/cm³; axis x: diameter of particles [nm])

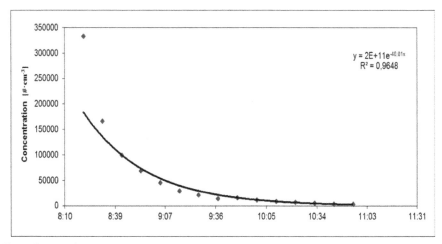

Figure 9. Rate of coagulation and deposition of nanoparticles in a workshop after welding plotted against time

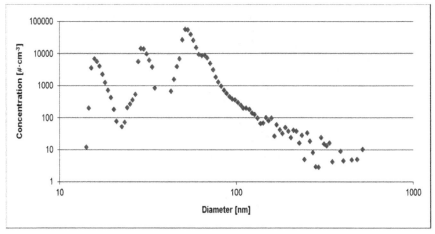

Figure 10. Revolver type weapon (King Cobra), two magazines were shot out with 6 bullets each

It was fairly alarming to find out that particles smaller than 50 nm were essentially most frequently present at a fire, extinguishing, welding, at an outlet from the modern Diesel engine and from a non-heated classic Diesel engine, and in large quantities after shooting.

Results for the Diesel engines are in agreement with the discussion and with the statement [4] that the improved combustion in modern Diesel engines extremely reduces the fraction of large particles; however, this is counterbalanced by the generation of extremely small particles: "No smoke coming from the exhaust pipe is reassuring to the eye, but the problem is in just that which cannot be seen."

3. Part II – Results of systematic measurements

3.1. Concentration and distribution of aerosol nanoparticles in the Prague subway station Muzeum C

In March 2011 an experiment was organized with the objective of measuring the distribution of size and concentration of aerosol particles in a very busy (changing) subway station in Prague. The location of the measurements was the station Muzeum C for a period of twelve hours (the data from the meteorological station were collected from 7:15 to 0:15, the measurements were performed from 7:40 to 0:28).

The measuring technology was situated in the middle of the platform (see Figure 11). The instrument enabled measurement in the range from 14.1 to 791 nm with a sampling interval of 5 minutes.

The trains on the line C are M1 (engine power 141.5 kW). The basic data about the platform dimensions are provided in Table 3.

Figure 11. Location of the measuring technology on a platform of the subway station Muzeum C

Length of the station	194 m
Depth of the platform center under the ground level	10 m
Platform width	10 m
Platform height	4.3 – 5 m

Table 3. Dimensions of the subway station Muzeum C

The temperature at the platform during the measurements was 12 – 13 °C, after 21:00 hours the temperature dropped by 1.5 – 2 °C, and the air flow at the measuring device had a pulsating character – see Figure 12.

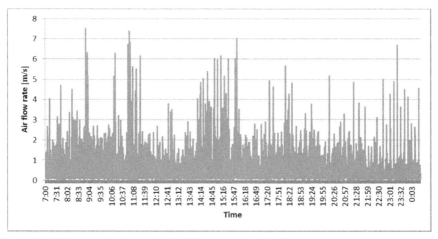

Figure 12. Air flow speed in the proximity of the measuring technology

The temperature outdoors in the morning hours was -4 °C, in the afternoon 2 °C, and in the evening 0 °C. The air flow on the surface was from 2 m.s^{-1} to 7 m.s^{-1} (the maximum was reached between 13:00 and 14:00 hours).

3.2. Results of the measurements and discussion

The basic results of measurements of the overall concentration of aerosol particles during subway operation are shown in the diagram in Figure 13. The diagram of the concentrations of aerosol particles is completed with the intervals of subway trains passing through the station.

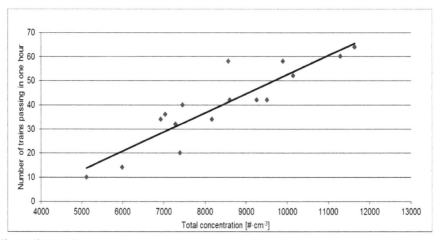

Figure 13. Overall concentration of aerosol particles during the subway operation

The diagram shown above indicates that the concentration of aerosol particles in the subway station is significantly affected by the frequency of trains passing through the station.

This fact has been confirmed by the following Figure 14, which presents a graphic rendering of the dependence of the overall concentration of aerosol particles on the number of trains passing through the subway station in both directions per hour. In theory, the diagram suggests that if there were no trains passing through the station the overall concentration of aerosols would be ca. 3400 particles per cm^3.

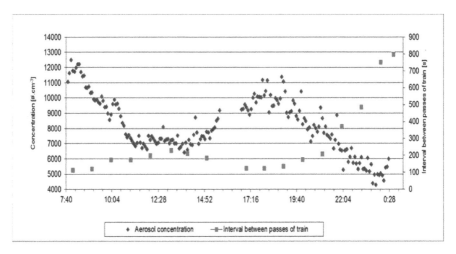

Figure 14. Comparison of the overall concentration of particles with the number of passing trains

The summary of average concentrations of aerosol particles for the monitored period is shown in Table 4. As a curiosity, we have provided also the measured concentration of aerosol particles in the environment during passage of a servicing Diesel locomotive MUV – 72 (engine TATRA T 928 – 2 with engine power 130 kW) after the passenger traffic in the station was closed.

	N.cm^{-3}	Standard deviation
Overall average (without the locomotive)	8200	±1800
Morning rush hour (7^{45} - 9^{45})	10800	± 950
Morning low (11^{30} - 14^{30})	7200	± 420
Afternoon rush hours (17^{30} - 19^{30})	9980	± 700
Night low (22^{00} - 0^{30})	5520	± 600
Maximum during the passage of the servicing locomotive	35500	-

Table 4. Average concentrations of aerosol particles during the monitored period

Figures 15 and 16 are graphic renderings of distribution of the particles during the morning and afternoon rush hours and during the traffic lows.

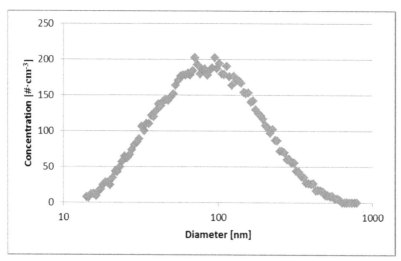

Figure 15. Distribution of particles during the morning and afternoon rush hours

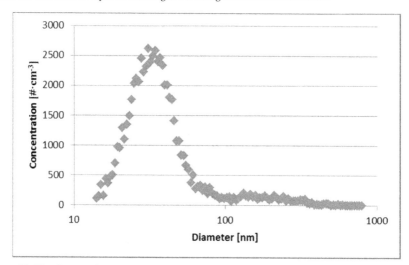

Figure 16. Distribution of particles during the morning low

Figure 17 is a graphic rendering of the distribution of particles during the passage of the servicing locomotive that passed through the station after it was closed for passenger traffic. The diagram indicates a distribution shift to smaller, i.e. more dangerous dimensions of the particles.

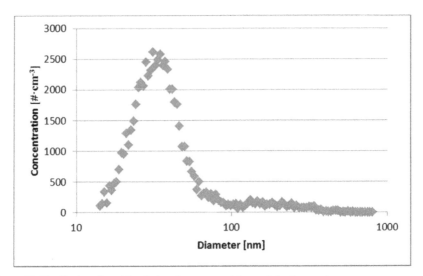

Figure 17. Distribution of particles during the passage of the servicing locomotive

What are the supposed sources of nanoparticles on the subway station platform?

a. Supply of the outdoor air via a vent shaft (see Figure 18) into the tunnel premises in the platform proximity: In winter the ventilation air is supplied into the subway from the ground level into the tunnel premises via a vent shaft. In this case the vent shaft is situated in a very close proximity to a busy road (Prague arterial road), where the level of traffic is a "slow moving traffic jam". The traffic on the arterial road probably influences the quantity of nanoparticles on the subway platform.

b. Release of previously deposited nanoparticles from the tunnel premises: The increased concentration of nanoparticles is probably also influenced by pressure waves caused by passing trains in the narrowed premises of the tunnels. These may be e.g. nanoparticles generated by wear of the tunnel lining (usually reinforced concrete), wear of the rails, crossties, subbase and by technical operations (servicing technology).

c. Braking of the trains: Another factor that may influence the measured quantity of nanoparticles is braking of the trains. The weight of a subway train is ca. 130 t and it brakes for several seconds. This braking results in the wear of wheels, brakes, rails, etc.

d. Influence of passengers: During the experiment we attempted to limit this influence to the maximum extent by placing the measuring technology at the end of the platform, but we still anticipate that the measured values might have been influenced by the changing numbers of persons in the station.

We anticipate that the mostly tunnel character of the subway line is the source of aerosol nanoparticles of various origin. Their spread (release) is probably caused by pressure and impulse waves created by the running subway trains.

Figure 18. The subway vent shaft structure on the ground level near the busy road

This assumption has been confirmed by a comparison of the quantity of measured nanoparticles inside a train car on the subway platform on the line C with the quantity measured in a subway train car during its trip on the line C (see part I). The concentration of nanoparticles in the travelling train car was higher than at the same place measured on the platform. This is probably caused by the ventilation system of the cars which takes in the air from the tunnel premises, plus by the higher concentration of passengers per area unit.

3.3. Aerosol and dust particles generated during processing of selected exotic woods

The objective of the measurements was to measure quantities and distribution of aerosol micro and nanoparticles generated by individual technological steps during processing of various types of tropic woods used on the market in the Czech Republic. At the same time, we also focused on the microstructure of the wood dust in the deposits and difference in the chemical composition of the individual woods; this may play a negative role after they get into the respiratory system or into contact with skin or eye mucosa. We focused on tropical woods due their wide variety and the dramatic increase of their import to the processing market in the Czech Republic.

Wood processing generates wood dust which may, depending on the size of the particles, form an aerosol or settle directly. The wood dust contains chemical substances that form the wood (polysaccharides, such as cellulose and hemicellulose, aromatic substances, such as lignin and tannins, resin terpenes, lipids, nitrogenous substances, inorganic substances etc.) depending on the wood condition, while it is impossible to exclude the presence of biological organisms, fungi, mildews or bacteria [5].

A negative effect of the wood dust on the human organism may occur in case of contact with skin or eye mucosa or inhalation by the respiratory tract. There is a general rule that the with decreasing size of the particles their respirability increases as well as their ability to bind with other substances (by sorption or condensation). Dusts from biologically highly aggressive woods may cause dermatitis, respiratory diseases, allergic respiratory problems (asthma) and carcinogenic effects (adenocarcinoma of nasal cavity and paranasal cavity). The chemical composition of wood opens a number of possibilities in contact with the biological system [6].

The measurements were conducted in the course of full operation in a production hall (area 700 m^2, volume 3 500 m^3) equipped with a state-of-the art filter and extraction system made by Cipres Filtr with the power output 37 kW, with a box filter CARM situated outside the hall. Despite this after a time the overall concentration of nanoparticles in the production hall increased to 4–5 x 10^5 N/cm^3. The difference from the initial level before the shift beginning is shown in the diagram in Figure 19.

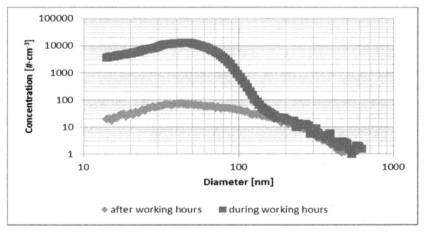

Figure 19. Comparison of the distribution of aerosol particles in the production hall during the operation and before the beginning of the operation

This has caused problems with identification of the technological process that generates the highest quantity of micro and nano aerosols. Despite those difficulties, we have identified the operation of the belt grinding machine as the main source of pollution in the production hall. The next experiment was conducted during the night, only with the technological operation of wood surface grinding with a grinding belt (belt grinding machine HOUFEK, PBH 300 B BASSEL, belt speed 17 m/s, grinding belt roughness AA 80, AA 100).

The temperature and humidity in the production hall: 24-25°C, 55%.

Tested woods: Ipé, Jatoba, Massaranduba, Merbau, Bangkirai, Faveira, Garapa, Teak, Bilinga.

The basic information on the tested tropic woods and on their processability and toxicity reported in literature is provided in Table 5.

Trade name of wood species	Latin name of wood species	Occurrence	Note about the wood processing [6]	Note about the toxicity [6]
Ipé	*Tabebuia spp.*	Central and South America	Planing is difficult, high-performance machinery is needed, very strong material, highly durable	Sawdust and grinding dust contain lapachol – as the dyestuff, it is irritating, may damage mucosa and cause dermal problems
Jatoba	*Hymenaea spp.*	Central and South America	Sawing – high-performance machinery is needed	Risk of mucosa and skin damage
Massaranduba	*Manilkara spp.*	South and tropical America (Brazil, Columbia)	Sawing – high-performance machinery is needed	Sawdust may be irritant, wood dust may irritate mucosa and skin
Merbau	*Intsia bakerie Prain*	Southeast Asia (Indonesia, Malaysia)	dulls the tools, difficult processing – special tools are needed	Chemical reaction with iron
Balau, Yellow (Bangkirai)	*Shorea argentea*	Southeast Asia (Malaysia, Indonesia)	Difficult processing	Not detected
Faveira	*Porkia spp.*	Tropical South America (Brazil, Columbia)	Easy sawing, processing without difficulties	Poor resistance against fungi and insects
Garapa*	*Apuleia Leiocarpa*	South America (Brazil)	Not detected	Allergenic and toxic
Teak	*Tectona Grandis*	Southeast Asia (Indonesia, Burma, Laos)	Sawing not completely easy, makes the tools blunt	Wood dust irritates skin, contains oily resins, resists decay
Bilinga (Opepe)	*Vaucle and Diderrichii*	West Africa (Sierra Leone, Nigeria, Cameroon)	Sawing requires high-performance machinery	Resistant against termites, the bark contains alkaloid

Table 5. Basic data about the tested woods

The layout of the measuring technology is shown in Figure 20.

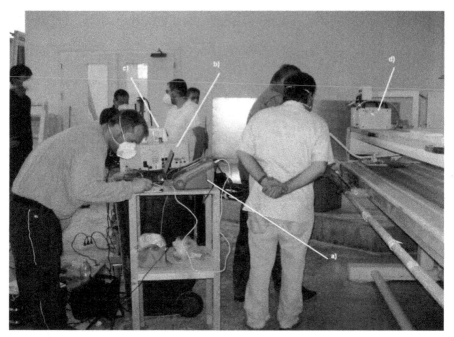

Figure 20. Layout of the measuring technology in respect to the belt grinding machine:

a. measurements of FIT factors,
b. measurements of distribution of aerosol particles (micro),
c. measurements of distribution of aerosol (nano) particles, d) cascade impactor.

In addition to the measurements of quantities and distribution of nano and micro aerosol particles we also measured the FIT factor to verify protective capacities of the respirators and collected samples of sedimented dust (sawdust).

3.4. Results and discussion

3.4.1. Determination of quantities of grinded-off wood

The weighted samples (mostly with the same area sizes) of exotic woods were grinded under the same conditions for 5 minutes on a belt grinding machine (see Figure 20). After the grinding was completed, the samples were weighed and the weight loss was converted into the area per 1 cm^2. The results shown in Table 6 indicate that the highest weight loss was found for the wood Garapa, while the values for Massaranduba, Ipé and Teak were comparable. The wood most durable against the employed grinding method was Merbau. We also compared the quantities of ground-off wood material depending on the grit size of the grinding belts and we found out that finer surface resulted in a higher weight of the ground-off material by up to 20-25% .

Wood (wood species)	Ground area (cm²)	Ground-off quantity (g)	Ground-off quantity per cm² (g)
Massaranduba	371	246	0.66
Ipé	371	261	0.70
Garapa	371	308	0.83
Teak	331	266	0.80
Bilinga	466	182	0.39
Jatoba	371	167	0.45
Faveira	371	157	0.42
Bangkirai	308	195	0.63
Merbau	371	78	0.21

Table 6. Determination of quantities of ground-off woods

3.4.2. Distribution of nanoparticles released during grinding of exotic woods

Examples of measured values of concentrations and distributions of aerosol particles in the range 15 – 750 nm generated by grinding of exotic woods after subtraction of the background are shown in the Figure 21.

The comparison of the above-presented diagrams of the distribution of nanoparticles in the range 7 – 100 nm has shown a detailed distribution of aerosol particles of the Ipé, Jatoba and Massaranduba woods (that belong to the category of harder materials), with the maximum at ca. 40 nm, while for the Merbau, Bangkirai and Faveira woods the maximum value shifted towards lower values.

The diagrams presented below (Figures 22 and 23) document that if we replace the grinding belt with a finer one the quantity of nanoparticles released into the atmosphere will increase, and the sizes of the particles will shift to lower values.

3.4.3. Analysis of sedimented dust:

The collected samples of sedimented dust after the wood grinding were subject to IR analysis, a microscopic study of the wood material and their thermal stability. The IR analysis sought to find a certain correlation between characteristic vibrations that may be related to the toxicity of the wood dust. The achieved microstructure of the dust was expected to provide information about the level of degradation of the wood structure by mechanical means (grinding). The thermal stability of the sedimented dust was expected to indicate the fire risks to be expected for the individual woods, while those results will be published separately.

Infrared spectroscopy is one of the few non-destructive methods for investigating the chemistry and physics of wood. Gradually, absorption bands with wave numbers have been defined that characterize the dominant building elements of the woods, such as cellulose, hemicellulose and lignin. We have also used the FT-IR (Fourier Transform Infrared) technology to measure infrared spectrums of the collected samples of sedimented dusts. Based on the published spectrums of similar woods and catalogue values of vibrations for the specific bonds and groups [7-10], we have made assignments to the individual absorption bands. As an example, we have made assignments to the measured values of the Massaranduba wood spectrum; see Table 7 and Figure 24.

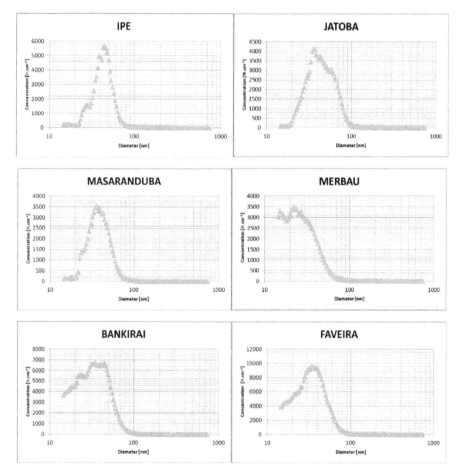

Figure 21. Distribution of aerosol particles of the woods Ipé, Jatoba, Massaranduba, Merbau, Bangkirai and Faveira, released during their grinding

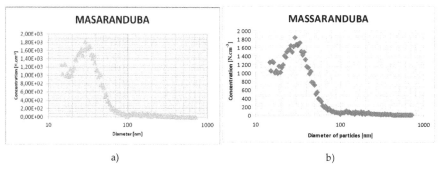

a) b)

Figure 22. Distribution of aerosol particles of the Massaranduba wood a) grain size 80, b) grain size 100 (Axis y: concentration of particles/cm³; axis x: diameter of particles [nm])

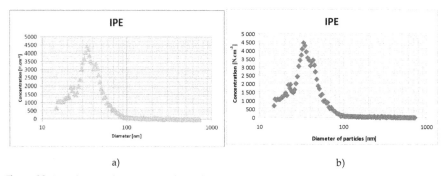

a) b)

Figure 23. Distribution of aerosol particles of the Ipé wood a) grain size 80, b) grain size 100

Vibrations (cm⁻¹)	Assignment of the functional group or skeleton
3343	O-H valence bond
2921	C-H valence bond in methyl group
1731	C=O ketone and in ester group
1593	Aromatic skeleton, valence bond C=O
1504	Aromatic skeleton, valence bond C=O
1454	C-H deformation asymmetric -CH₃ and –CH₂–
1422	Vibration in the aromatic skeleton by combination with the deformation vibration in the C-H plane
1368	C-H deformation vibration in cellulose and hemicelluloses
1317	C-H vibration in cellulose and C-O vibration in syringyl derivatives
1232	Syringyl[a] skeleton and bond vibration C= in lignin and xylan[b]
1155	C-O-C vibration in cellulose and hemicelluloses
1023	Aromatic C-H deformation in the plane, C-OH, C-O deformation
894	Glycoside bonds
814	Planar vibration of the mannose ring

Table 7. Assignment of the wave numbers of absorption belts to specific groups or skeleton of the Massaranduba wood material. [a] As a part of lignin, [b] As a part of hemicellulose

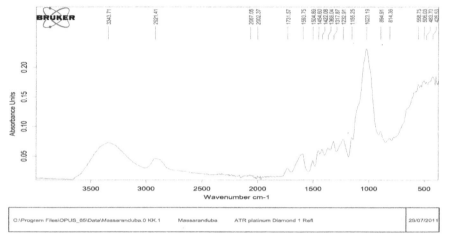

Figure 24. IR Spectrum of sedimented dust from Massaranduba wood

The comparison of the measured spectrums has shown a relative identity, particularly outside the fingerprint area of the molecule.

The variance in the wave number of C-OH vibrations is 15 cm^{-1} between the individual woods (the highest wave number 3350 cm^{-1} is for Ipé and Jatoba, the lowest is 3335 cm^{-1} for Bankirai). The variance of the valence bond vibration C-H is 72 cm^{-1} (the highest wave number is 2921 cm^{-1} for Massaranduba, the lowest is 2849 cm^{-1} for Bangkirai). The vibration shifts are probably caused by intermolecular hydrogen bonds that affect the wood density.

In the molecule fingerprint area, we focused on the identification of characteristic vibrations for various lignin skeletons and chinoid bonds (lapachol). The differences between the spectrums of the individual woods were demonstrated by absorbance values. Before the microscopic examination of the samples of sedimented dust (we will hereinafter use the term sawdust to refer to its method of origin) we described some of its external macroscopic properties and they were later confirmed at microscopic magnification by the factor of 40; they may be briefly described as follows:

- Ipé: fine sawdust with minimum dustiness
- Jatoba: sawdust of the same size as Ipé but, unlike Ipé, with significant dustiness
- Massaranduba: coarse sawdust with a structure similar to oak
- Merbau: the finest sawdust structure, with the highest dustiness
- Bangkirai: medium sawdust with a significant representation of finer particles
- Faveira: similar structure and distribution of fractions as Bangkirai
- Garapa: fine sawdust with medium dustiness and good powderiness
- Teak: sawdust with coarse structure and minimum dustiness, caused by cohesiveness or aggregation of sawdust
- Bilinga: coarse structure of sawdust with medium values of dustiness and powderiness

A summary overview of identifiable macroscopic properties is shown in Table 8

	Dustiness	Inherent cohesiveness	Powderiness	Size of particles	Structure
Ipé	+	++	+	++	Coarse
Jatoba	++	+	++	++	Coarse
Massaranduba	++	+++	++	+++	Coarse
Merbau	+++	+	+++	+	Fine
Bangkirai	+	++	++	++	Medium coarse
Faveira	+	+++	+	++	Medium fine
Garapa	++	+	++	+	Coarse
Teak	+	+++	+	+++	Coarse
Bilinga	++	++	++	++	Coarse

Table 8. Description of macroscopic properties of sedimented particles

We were interested in the shape of the particles, which will probably play a role in their fixation in the respiratory tract, so we made microscopic pictures. As an example shown below, we have provided microscopic pictures magnified 200 times, while the line segment on the pictures represents 100 micrometers; see Figure 25.

Other risks of nano-, micro- and dust particles are physicochemical, i.e. risk of fire, explosion, uncontrolled and undesired reaction. For this reason, the samples of sedimented dust were subject to thermal gravimetric analysis. For all samples of sedimented dust generated by coarse grinding, the thermal decomposition resulted in two separate exothermic processes T_1 in the range 279-333 °C (the lowest for Merbau) and T_2 in the range 402 -437 °C (the lowest for Garapa). After summarizing thermal processes during thermal decomposition of dusts of our woods, the highest thermal effects were found for the woods Jatoba (5904 kJ/kg) and Garapa (5506 kJ/kg), while the lowest value was found for the wood Teak (2210 kJ/kg).

Another finding with a safety impact was that if we use finer grains for the grinding of some woods, e.g. Massaranduba, the exothermic effect is much less significant – see the diagrams in Figure 26.

We can thus conclude that, despite a modern extraction system that was installed in the workshop, the content of aerosol nanoparticles was two orders of magnitude higher than before the production works started, and the concentrations of dust particles in the immediate proximity of the grinder were several times higher than values permitted by Czech legislation. The sizes of aerosol nanoparticles, based on the determined distribution, mean that they can pass through protective barriers of the respiratory system up to the alveoli. Here the question remains on the role played in the toxicity by the concentration, chemical composition, surface and shape of the nanoparticles.

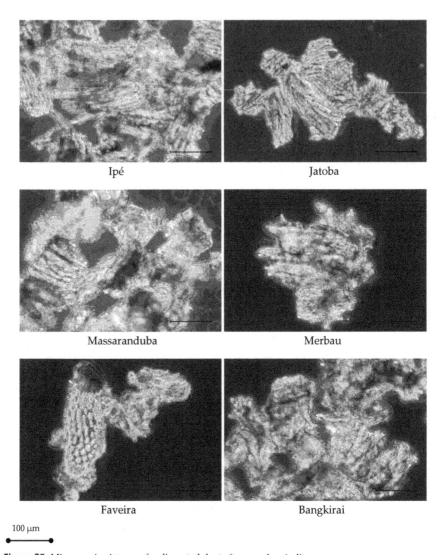

Ipé   Jatoba

Massaranduba   Merbau

Faveira   Bangkirai

100 μm

Figure 25. Microscopic pictures of sedimented dust after woods grinding

Apart from the dominant polymeric components of the woods (cellulose, hemicellulose, lignin), the woods also contain low-molecular substances. Those substances are sometimes classified as so-called extractable components, and they can be extracted from the wood material by various combinations of extraction agents.

Many of those substances, such as terpenoids, phenols, tannins, chinons, stylbens, flavonoids, alkaloids, etc., feature biological activity, both positive and negative. The disruption of wood matter by technological operations means that one can anticipate different distributions of dominant wood components as well as low-molecular substances on the surfaces of nanoparticles (microparticles).

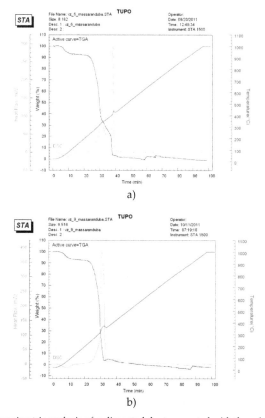

a)

b)

Figure 26. Thermal gravimetric analysis of sedimented dust generated with the grinding belt grain size 80 (a) and 100 (b)

The differences in the size, shape, and certain physical properties of the particles of sedimented dust from the wood grinding have been described above. We can only speculate about the extent to which the shape of particles may influence their toxic effects. Sharp particles may behave in the organism similarly as has been described for asbestos (pulmonary fibrosis) or chronic tracheitis.

The size of the particles influences their shape, which we have illustrated with the shapes of particles generated by grinding Massaranduba and Jatoba woods. Figure 27 shows a comparison of microscopic pictures of particles with the size of hundreds of µm and particles with the size of units of µm, made by electron microscope, using particles trapped between the levels A and B in the cascade impactor.

Figure 28 shows electron microscope images of particles generated by grinding the Massaranduba wood trapped between A-C sorting levels in the cascade impactor for the Jatoba wood.

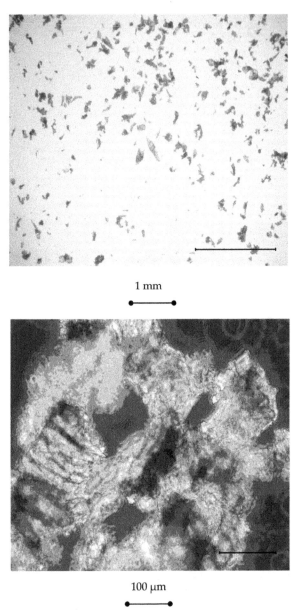

1 mm

100 μm

Figure 27. Comparison of the shape and size of Massaranduba wood particles

A general conclusion can be drawn that the world's major occupational health agencies only provide warnings about the risks in their reports for selected individual types of wood, and they request better protection of particular body parts (skin, eye mucosa, respiratory tract).

Our measurements have led to a recommendation that the selection of safety measures to protect the health of the employees during some technological operations with woods should take into account the character (type) of the processed wood.

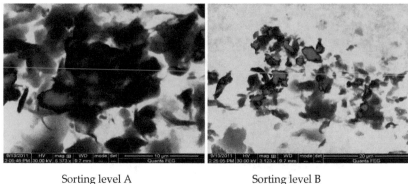

Sorting level A   Sorting level B

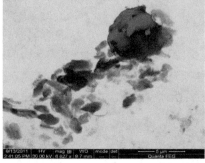

Sorting level C

Figure 28. Electron microscope pictures of Jatoba wood particles generated by grinding.

Based on the available toxicological information about a particular wood, or based on measurements similar to those we have made for the anticipated technological operation, it will be necessary to specify or to expand the preventive measures with the objective of minimizing the contact of workers with particles generated by the processing. The measures may include technical (e.g. wetting system) and organizational measures (e.g. shorter exposure, alternation of workers), personal protective equipment (e.g. HEPA respirators: our measurements have demonstrated their 93-97% effectiveness) or health related measures (e.g. shortened intervals between medical checkups).

3.5. Measurements of quantities and distribution of aerosol nanoparticles in some steelworks operations

Metallurgic operations rank among the biggest producers of wastes of all types and categories. They include e.g. production wastes, slag, waste sludge, wastewater and emissions of heavy metals, associated with high-temperature processes of metal vapours formation and their condensation or potential chemical transformation.

Thanks to the pro-active approach of the company EVRAZ Vítkovice Steel, a.s. to the environment and safety of their employees, we were able to perform measurements of quantities and distribution of aerosols during the operation in various parts of the steelworks that utilize oxygen steelmaking.

The basis of the oxygen steelmaking (i.e. oxidation) is the removal of undesired impurities from the raw iron melt. The key elements that are converted into oxides in the process are carbon, silicon, manganese, phosphor and sulphur.

The oxygen steelmaking process is discontinuous and may be divided into the following steps:

a. Preparation and storage of metal melt
b. Pre-treatment of the metal melt (desulfurization of the melt by introduction of calcium carbide, magnesium and lime)
c. Oxidation in the oxygen converter
d. Secondary metallurgy (i.e. vacuum metallurgy)
e. Casting (slab casting)

Before the experimental measurement, we attempted to identify locations with the expected increased emission levels, specifically:

- at the converter gas,
- during the desulfurization process,
- during handling of scraps and iron ore,
- during slag removal,
- during casting, pourover of raw iron or steel.

Meanwhile, we had to consider the safety of the workers performing the measurements and sensitivity of the employed technology to the environment in which it operates, e.g. high temperature, explosive environment, etc.

The selected locations in the premises of continual steel casting and in the converter hall of the steelworks represented the resulting compromise.

3.6. Experimental part

Measurements were conducted under regular operation of the steelworks. For safety reasons, measuring instruments to measure quantities and distribution of aerosol nanoparticles were situated only in the following locations:

a. at the equipment for continual casting, ca. 3 m from the slab, which had been already in the horizontal position and in the area of the so-called secondary cooling, ca. 6 m from the flame-cutting machine (Figure 29).
b. in the steelworks dispatching section, ca. 3 m from the scarfing machine, where the cooled slab was parted crosswise (Figure 30).

Figure 29. Measuring point at the slab continual casting

Figure 30. Measuring point in the steelworks dispatching section (at the scarfing machine)

Measurements of particles trapped in personal cascade impactors were performed in the convertor hall of the steelworks in 2 selected locations under the technological conditions described below.

3.7. Distribution of nanoparticles in the premises of the continual casting equipment

3.7.1. Measuring point a)

The average flow rate in the first location was 0.12 m·s⁻¹ (determined with TESTO 445 with a thermal probe). 6 spectrums were measured in total with the distribution of size of aerosol particles ranging from 14 to 736 nm. The distribution of the size of aerosol particles obtained by averaging the collected spectrums is shown in the diagram in Figure 31, Table 9.

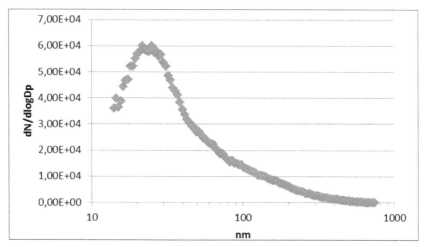

Figure 31. Distribution of aerosol particles in the measuring location a)

The mode of the collected spectrum is around 20 nm. The presented spectrums indicated presence of particles under 10 nm, which may be estimated from the size distribution

Measuring point a)	Overall concentration (N·cm⁻³)	Total surface of the particles (nm²·cm⁻³)	Total volume of the particles (nm³·cm⁻³)	Total weight of the particles (µg·m⁻³)
Average values	3.57 . 10⁴	6.46 . 10⁸	2.33 . 10¹⁰	28

Table 9. Measured physical values of nanoparticles – point a)

3.7.2. Measuring point b)

The average flow rate in the location was 0.27 m·s⁻¹. The distribution of size of aerosol particles obtained in this measurement location is shown in the diagram in Figure 32, Table 10.

The higher flow rate has probably also affected the uneven distribution of the size of aerosol particles. The different technological development of the operation during which the measurement was performed also played a role.

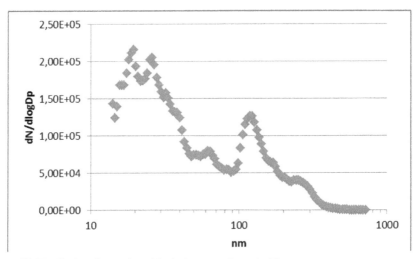

Figure 32. Distribution of aerosol particles in the measuring point b)

The maximum peak of the doublet shape in the area around 20 nm corresponds to the distribution measured in the measuring point a) but the concentration of the particles was higher. Another area with an increased number of particles is around 120 nm, while this phenomenon was not observed in the previous case. The overall concentration of particles in the nano area is by one order of magnitude higher, the weight of the particles 5x higher than in the previous case.

Measuring point b)	Overall concentration $(N \cdot cm^{-3})$	Total surface of the particles $(nm^2 \cdot cm^{-3})$	Total volume of the particles $(nm^3 \cdot cm^{-3})$	Total weight of the particles $(\mu g \cdot m^{-3})$
Average values	$1.4 \cdot 10^5$	$4.03 \cdot 10^9$	$1.38 \cdot 10^{11}$	166

Table 10. Measured physical values of nanoparticles – point b)

3.8. Measurements of the weight of trapped aerosol particles in selected locations of the steelworks convertor hall

During the performed technological operations in the steelworks convertor section, the following measuring points were selected:

- during the pourover of pre-treated raw iron into the converter after desulfurization,
- during the pourover of pre-treated raw iron from the railway carriage at the blast furnace into the ladle,

with impactors for collection of aerosol particles sorted in agreement with the specified sorting levels (Table 11). The table contains the subsequently weighed particles by fractions and conversions into volume concentrations in the proximity of the technology operation.

Sampling point	Sorting level (μm)	Trapped particles a) (mg)	Conversion into volume concentration (mg/m³)
Pourover into the pre-treated melt	A 2.5	0.12	2.40
	B 1.0	0.15	3.00
	C 0.5	0.09	1.80
	D 0.25	0.12	2.40
	< 0.25	0.18	3.60
	Total	0.66	13.20
Pourover of raw iron	A 2.5	0.38	7.60
	B 1.0	0.35	7.00
	C 0.5	0.45	9.00
	D 0.25	0.33	6.60
	< 0.25	0.38	7.60
	Total	1.89	37.80

a) air flow rate 10l/min. for a period of 5 minutes.

Table 11. Individual trapped fractions of aerosols in the first impactor

Samples of trapped particles at the pourover of raw iron were submitted for electron microscope analysis. The images made by the electron microscope based on the particles trapped in the filter by sorting levels are shown in Figures 33 and 34.

Pictures made by a scanning electron microscope show visible particles from several hundreds of nanometers to ca. 5μm. The evaluation of the sizes of the observed particles has made it possible to estimate that the most numerous particles were in the range from 1 to 2μm. The prevailing majority of trapped particles in fine atmospheric aerosols were spherical; see Figure 33. In agreement with the generally accepted theory, particles of the size of units of micrometers are formed directly by the solidification of finely dispersed liquid aerosol of liquid iron. If the cooling rate is sufficient, round particles with some signs of crystalline structure of atoms on the surface appear instead of crystalline formations. On the contrary, if the conditions for a transition into a solid state are different, particularly in terms of the cooling rate, then fairly interesting crystalline formations can be found between the particles, as shown in Figure 34. The resulting product is actually an aggregate of very small crystals which came into immediate contact in the atmosphere.

The entire process of formation of the fine aerosol is accompanied by a chemical reaction in which melted iron particles are in a thermodynamic imbalance with oxygen from the atmosphere, and therefore an intense exothermic chemical reaction occurs on the surface of the particles which produces oxidation products of iron. The resulting formations are shown in the pictures. A chemical analysis of particles with EDS has confirmed the variable values of the Fe/O ratio. In some cases the atomic ratio Fe/O was > 3/2, which may be explained by the presence of non-reacted iron in the particle core.

Sorting level A   Sorting level B

Sorting level C   Sorting level D

Figure 33. Images from electron microscope of nano- and microparticles from the environment of the pourover of raw iron from the railway carriage

Samples of particles trapped in the impactor were collected at the site where pre-treated melt of raw iron was poured over after desulfurization, and they were analyzed by electron microscope. In addition to the minor number of spherical particles (composed of iron, iron oxide and iron-calcium), the images also show carbon-based non-spherical and non-metallic particles. A characteristic illustration of particles trapped in the impactor is shown in Figure 35. The content of the trapped particles was probably influenced by the composition of the residual slag that remained in the melt after the tapping. The melt desulfurization, as mentioned at the beginning, is performed by the addition of calcium carbide, magnesium and lime, so the presence of calcium is completely logical.

Sorting level A   Sorting level A

Figure 34. Images from electron microscope of trapped particles of non- typical shapes from the environment of raw iron casting

Figure 35. Electron snapshots of trapped nano- and micro- particles from the site of the pourover of pre-treated melt after desulfurization

4. Discussion about the toxicity of iron particles

Iron in suitable concentrations is an element essential for human health which participates in the transport of oxygen (hemoglobin, myoglobine) in cellular breathing. If the concentration of iron in the organism exceeds the capacity of transport and spare proteins for iron, then it is deposited loose into the tissue. An increased deposition of iron in tissues causes fibrotization and the reduction of functional tissue. The main signs include development of liver hemosiderosis and later cirrhosis, type II diabetes, cardiomyophy, arthritis etc. [11]

The round shape of nanoparticles probably reduces toxicity and facilitates transport from alveoli by breathing.

The concentrations of micro- and nano- particles measured in one cubic meter were higher than the permitted exposure limit (PEL) specified in the Government Order of the Czech Republic but, considering the mostly automated operation of the steelworks, the level of risk to the employees is not significant.

5. Conclusion – Examples of measures recommended based on the results of the measurements

-    When designing new subway routes, the environment of the vent shaft outlets on the ground level should be taken into account
-    Install effective filters into the vent shafts
-    Perform consistent cleaning of the subway lining
-    Replace the subway fleet (electric drives)
-    Policemen should wear respirators and goggles for shooting practice
-    Limit the use of entertainment pyrotechnics in residential areas
-    Get as much information as possible about the origin and toxicity before processing any exotic woods
-    Wear HEPA-respirators and goggles when processing exotic woods and working in the steelworks.

6. Instrumentation

The *measurements of aerosol particles* were conducted with SMPS (Scanning Mobility Particle Sizer) 3934 consisting of CPC (Condensation Particle Counter) 3022 (working in low regime) and EC (Electrostatic Classifier) 3071 equipped with DMA (Differential Mobility Analyzer), probe 3081 and impactor 0.0457 cm. The measurements were conducted using the setup scanning interval of 5 minutes, which enabled collection of samples in the range from 14 nm to 740 nm.

Aerosol particles were collected by means of a personal cascade impactor (Sioutas 225-370). The personal cascade impactor consists of four impaction levels and a filter that enables sorting and sampling airborne particles in five size intervals. Samples were collected with the personal impactor which used the sampling device QuickTake30 at a specified constant flow rate.

Micro *aerosol particles* were measured in the range from 0.5 μm to 20 μm using APS (Aerodynamic Particle Sizer Spectrometer 3321 TSI USA).

Measurements of FIT factors of the protective respirators, device PortaCount Pro+ 8038 TSI, method MAZL – 40/11 (SÚJCHBO), OSHA 29CFR1910.134.

Thermal analysis, TG-DSC and TGA were measured with the device STAi 1500 made by Instrument Specialists Incorporated - THASS

Infrared ATR spectrums (FTIR) were measured with the spectrometer Nicolet 7600 (Thermo Nicolet Instruments Co., Madison, USA) with the detector DTGS and beam divider KBr. Parameters of the measurements: number of spectrum accumulations 128, resolution 2 cm^{-1}. The measurements were conducted with an ATR cuvette Smart Orbit (Thermo Scientific) equipped with a diamond crystal.

Microscope: Microscope Olympus XI71 + CCD camera Olympus DP72 (Olympus Co., Japan) were used for magnification 200-600x.

El. microscope: made by FEI, model Quanta 450 FEG, high fiber method, ETD detector (secondary electrons) and BSE (back-scattered electrons).

Thermogravimetric and differential thermal analysis (GTA/DTA) was performed with STA; 1500 made by Instrument Specialists Incorporated – THASS.

Author details

Karel Klouda
State Office for Nuclear Safety, Czech Republic, e-mail: karel.klouda@sujb.cz

Stanislav Brádka and Petr Otáhal
The National Institute for Nuclear, Chemical and Biological Protection

7. Acknowledgement

Bílek K., Cejpek J., Dropa T., Kollárová D., Kubátová H., Matheisová H., Němeček Vl., Urban M., Večerková J., Vošáhlík J., Weisheitlová M., Witkovská V.

8. References

[1] Nohavica D. (2009) Respiratory and Cardiovascular Problems Related to Nanoparticles (in Czech). Proc. of Nanocon 2009, Roznov pod Radhostem, p. 76, ISBN 978-80-87294-12-3

[2] Klouda K., Kubátová H., Zemanová E. (2011) Nanomaterials: Prost an Contras. Communications 2/11, Vol. 13, pp. 1-13

[3] Klouda K. (2011) Analysis of results of Measurement of Nanoparticles from Antrhopogenis Sources (in Czech). Spektrum 1/2011, pp. 19-23

[4] Glatz A. (2010) Efficiency of Emission Particles Filters and Modern Engines Ecology are strongly Arguable (in Czech). [2010-03-16], Available: http://biom.cz/odborne-clannky/efektivita-filtru..

[5] Jankovský M., Lachman J., Staszková L. (1999) Wood Chemistry (in Czech). Czech Univerzity of Life Sciences. Praha, ISBN 80-213-0559-2

[6] Roček I. (2005) Wood of Tropical Areas (in Czech). Czech Univerzity of Life Sciences, ISBN 80-213-1346-3

[7] Zhou G., Taylor G., Polle A. (2011) Ftir-atr-based prediction and modeling of lignin and energy contents reveals independent infra-specific variation of these traits in bio energy poplars, Plant Methods, 7:9, pp. 1-10

[8] Jungnikl K., Paris O., Fratzl P., Burgert I. (2008) The implication of chemical extraction treatments on the cell wall nanostructure of soft wood, Celulose 15, pp. 407-418

[9] Lau S. (1992) FT-IR spectroscopic studies on lignin from some tropical wood and rattan, Pertanika 14, pp. 15-81

[10] Coates J. (2000) Interpretaion of infrared spectra and practical approach. Encyclopedia of Analytical Chemistry, John Wiley&Sons, pp. 10815-10837

[11] Houdak J. (2011) New information about iron metabolism in human body. Available: http://www.slideshare.net/jirihouda/metabolismus-eleza%20/ (in Czech)

Chapter 9

Investigation of Suspended and Settled Particulate Matter in Indoor Air

Adriana Estokova and Nadezda Stevulova

Additional information is available at the end of the chapter

1. Introduction

Particulate matter is a natural part of the atmosphere, where the solid or liquid particles are suspended in the air. These suspended particles, also known as suspended particulate matter represents a dispersion aerosol system. In the air there are many types of microscopic airborne particles originated from both natural and anthropogenic processes, such as atmospheric clouds of water droplets, photochemically generated particles, re-suspended particulates, fumes arising from the production of energy, etc. They are present in various forms, eg. mists, fumes, dust. The atmosphere contains particles of the size ranging from slightly larger than molecules up to hundreds of micrometers, which consists of a variety of chemical compounds [1]. Depending of their lifetime, the particulates observed at a location can be both of local origin or the product of the transport over distances of hundreds to thousands kilometres.

Particulate matter is mainly classified by particle size distribution as follows [2]: Coarse Particles (CP) include all particles with an aerodynamic diameter (diameter of a sphere with unit density and mass equal to the mass of the provided particle) greater than 2.5 micrometers and less than 10 micrometers. These particles are identified as $PM_{2.5-10}$. PM_{10} is an abbreviation used for so called „thoracic" particles with the diameter under 10 μm. Fine Particles (FP) include all particles having an aerodynamic diameter less than 2.5 micrometers and greater than 0.1 micrometers ($PM_{2.5}$). Ultrafine Particles (UFP) include all particles the aerodynamic diameter of which is less than 0.1 micrometers. These size limits are not sharp; the cyclone and impactor pre-separators remove half of the particles at the cut size and larger particles with increasing efficiency.

Increase in particulate matter air contamination and its negative impact on human health have resulted in efforts to monitor and identify the pollutants. The particulate mass concentrations in a very clean urban environment are about 10 μg.m^{-3}, which correspond to

2.10^7 particles in 1 m^3. In the polluted urban air the particle concentrations are higher than 10^{11} particles in 1 m^3 and their mass concentrations may be higher than 100 $\mu g.m^{-3}$ [1,3]. In the Slovak Republic, the average annual outdoor PM10 concentrations ranged from 11.6 –18 $\mu g.m^{-3}$ in 2009 [4].

Danger of toxic inhalation exposure depends on both the physical and chemical characteristics of particulate matter and thus the study of its properties is essential to assess the health risks. Exposure to PM in ambient air has been linked to a number of different health outcomes, ranging from modest transient changes in the respiratory tract and impaired pulmonary function, through increased risk of symptoms requiring emergency room or hospital treatment, to increased risk of death from cardiovascular and respiratory diseases or lung cancer. The elderly, children, and people with chronic lung disease, influenza, or asthma, are especially sensitive to the effects of particulate matter [5]. Multiple studies have showed that a short-term exposure to particulate matter may associated with increased cardiovascular mortality [6-8]. The occurrence of particulate matters in the air interferes with human health not only due to its composition but also due to its specific properties. The large specific particle surface takes a share on the catalysis of heterogeneous chemical reactions and on adsorption of other pollutants and their transport [9].

Sources of particulate matter occur in the outdoor air as well as in the indoor environment. Ambient air concentrations are strongly dependent on meteorological factors in contrast to the indoor environment which is much more stable. The suspended particulate matter present in the indoor air is cumulated and as reported by [10-12] the indoor particulate concentrations are often measured to be higher than those outdoors. With the emphasis on both energy conservation and efficiency, mainly new home construction can create the problem of indoor air pollution. Vapour barriers, tight windows, weather-stripping and caulk have reduced or stopped fresh air from infiltrating and replacing stale air. Special attention must be paid to indoor air contamination because people spend a substantial portion of their time in indoor environment [13].

If indoor air pollution is investigated, both outdoor and indoor sources have to be considered, because the outdoor air is an important source of indoor particles pollution. Indoor particle concentration depends on penetration of outdoor particles into the indoor environment and on intensity of indoor aerosol sources [2]. Indoor particulate matter sources include building materials, cooking, heating and all activities related to combustion processes, smoking, cleaning and moving of inhabitants [14,15]. The importance of indoor sources depends significantly also on the number and habits of the inhabitants. It was noted [16] that the concentration of PM2.5 was 2.8 times higher in houses where people smoked.

The behaviour of indoor aerosols is affected by the structural system of a building, material characteristics, the way of air exchange, the operating mode of indoor environment in the presence of inhabitants. The structural systems of a building along with the physical properties of the outdoor air (wind direction and intensity, the difference in the density of the indoor/outdoor air, the difference in the indoor/outdoor air temperatures etc.) determine interzonal transport of pollutants [17]. In multi-floor buildings, the flow induced by buoyancy influences the motion of contaminated air within the building. Mechanical and/or natural ventilation and infiltration define air exchange rate, and thereby the amount of outdoor particles penetrating into the building interior. The efficiency of filters integrated in mechanical ventilation systems and natural ventilation by open windows allows the estimations of particle penetration in the dependence on outdoor aerosol concentration, whereas infiltration through cracks in the building envelope is uncontrolled and depends not only on physical properties of contaminated air but mainly on particle deposition on surface cracks [18,19].

Operation, the number and behaviour of inhabitants, i.e. type, emission intensity and amount of indoor contamination sources determine temporal and spatial variations of indoor aerosol distribution. In addition, wet processes such as cleaning, washing, drying and ironing increase relative humidity which can lead to variations in particle size distribution [20]. Physical properties of employed building materials such as thermal conductivity influence surface-to-air temperature difference, thermal convection and thermophoresis (or thermoprecipitation). This process is significant in the winter season when constructions separate heated from unheated areas. Chemical composition of particulate matter can influence the appearance of the electrostatic charge. The total aerosol concentration is determined by the balance between source emissions and aerosol decay due to indoor air chemical processes and aerosol loss mechanisms [2].

This chapter aims to present the results of the investigation of both suspended and settled particulate matter occurring indoors. The mass concentration and surface concentration measured were monitored for suspended and settled particulate matter, respectively. The chemical composition with special regard to the metals content as well as the morphology of indoor particulates was studied.

2. Indoor particulate matter decay

The aerosol particulate decay in indoor environment occurs by two mechanisms - ventilation and deposition. In general, ventilation is a positive mechanism for the loss of particles from indoor air. However, in real conditions, it often may cause entering the outdoor pollutants with supplied air into the indoor environment. The extent which ventilation contributes to the reduction of the indoor concentrations depends on the way of air exchange which can be carried out by natural air change, infiltration or ventilation systems. If the ratio of indoor and outdoor concentrations I/O reaches a value more than 1, the positive venting mechanism will result in a reduction of particulate matter concentration due to dilution. Otherwise, the contamination of indoor air increases by addition of outdoor particulate matter, mainly by natural air change. Ventilation systems should ensure the particulate matter concentration in the indoor environment is not increasing due to utilization of special filters in the inlet. In addition, coarse particles in ventilation system are often deposited by gravitational process which also leads to the removing of particles from the air supplied. On the other hand, particles deposited in the pipes can be re-suspended in dependence on the air flow speed [21].

Particle deposition is an important factor affecting indoor particle concentrations in all types of buildings and is considered to be a dominant mechanism of the aerosols concentration level decreasing [22-23]. The largest incidental losses occur as a result of particle deposition onto the surfaces. Due to the relatively large surface-to-volume ratio indoors, deposition has a much larger effect on reducing concentrations indoors than it does outdoors [19].

Particle deposition on indoor surfaces strongly depends on particle size and is governed by the processes of particle diffusion toward the surfaces, which is of particular significance for very small particles, and of gravitational sedimentation, which is significant for larger particles. In addition, the presence of airflows induced by convection currents or the action of fans, as well as air turbulence, can increase particle transport towards the surface a thus the deposition. Deposition is also dependent on the surface area and on its characteristics, with sticky surfaces resulting in higher deposition than smooth one. The larger surface area, the higher probability of particle deposition, and therefore furnished rooms, with lots of surface area, will have a higher deposition rate than bare rooms. Additional factors affecting particles deposition are: the presence of surface charge, which leads to the deposition rate increasing; temperature gradient, resulting in convective currents and thermophoretic deposition; and room volume [2].

Aerosol particles adhere when they collide with a surface. The aerosol concentration at the surface is zero and the concentration gradient is established in the region near the surface. The concentration gradient causes a continuous diffusion of aerosol particles to the surface, which leads to a gradual decay in concentration. Applying Fick´s first law of diffusion, deposition rate J is defined as a number of particles depositing per unit surface area per unit time and is given by equation (1)

$$J = n_0 \left(\frac{D}{\pi t} \right)^{1/2} \tag{1}$$

where n_0 is the uniform initial concentration and D is the particle diffusion coefficient [12]. The deposition can be also characterized in terms of deposition velocity V_{dep}, which is defined as the deposition rate divided by concentration in the equation (2)

$$V_{dep} = \frac{J}{n_0} = \frac{number\ deposited\ /\ m^2.s}{number\ /\ m^3} = m\ /\ s \tag{2}$$

The number of particles depositing on the total surface per unit time is expressed by the deposition loss rate coefficient β [1/s, 1/h]. This coefficient includes all the processes that remove the particle in enclosure (e.g. diffusion loss, gravitational settling loss and other loss mechanisms by external forces). In the context of regular geometry, β can be evaluated from the deposition velocity on different orientation of surfaces and their particular surface area, and can be expressed as

$$\beta = \frac{V_{dw} A_w + V_{du} A_h + V_{dd} A_d}{V} \tag{3}$$

where A_w, A_h and A_d are the total areas for the vertical wall, upward-facing and downward-facing horizontal surfaces, respectively. V_{dw}, V_{du} and V_{dd} are the deposition velocities for the vertical wall, upward-facing and downward-facing horizontal surfaces, respectively, and V is the volume of the enclosure [13].

Diffusion deposition is primarily observed on vertical and downward-facing surfaces (ceilings). Deposition induced by gravitational force is observed onto upward-facing surfaces (wear layer of floor constructions, upward-facing areas of furnishing). Air drag force compared with settling particle is determined by airflow. For settling observed in still air (i.e. Re < 1 laminar airflow) the Stoke's low is valid. If airflow is turbulent (Re > 1000), Newton resistant low is valid for settling particle. Terminal settling velocity VTS of the particle settling due to gravitational force is results of balance drag and gravity. V_{TS} is expressed in equations (4, 5) [1].

$$V_{TS} = \frac{\rho_p d_p^2 g}{18\eta}, \text{ for Re } < 1 \text{ laminar airflow} \tag{4}$$

$$V_{TS} = \left(\frac{4 \rho_p d_p g}{3 C_D \rho_g} \right)^{1/2}, \text{ for Re } > 1000 \text{ turbulent airflow} \tag{5}$$

where η is the viscosity of the air, ρ_p a ρ_g are the density of the particle and the density of the air, d_p is the particle diameter, g is the gravitational acceleration and C_D the drag coefficient. Indoor particle deposition can be induced also by thermophoretic forces which results in thermoprecipitation, or by ventilation and air conditioning use which lead to the eddy diffusion. Thermoprecipitation may be significant in the winter season because of heating. The presence of a heating device seems to be related to lower concentrations of a number of components, such as particle mass, Cr, Zn, Ca^{2+}, SO_4^{2-} and NO_3^- and other as noted in reference [45].

Particles deposited on indoor surfaces create a potential reservoir from where they can be re-suspended whereby the secondary contamination is increased. This re-suspension effect can be caused by mechanical vibration, aerodynamic or electrostatic forces.

3. Indoor air monitoring – A case study

The monitoring of aerosol particulate matter (PM) was carried out in three rooms of the selected flat building in the city of Košice, Slovakia. Kitchen, living room and working room as representative indoor environments with different indoor sources were chosen for PM monitoring. Environmental tobacco smoke was considered a major source of the particles in the living room; cooking on the gas stove was considered a major indoor source of particulate matter in the kitchen. None significant indoor source was identified in the working room. However, a penetration of outdoor particles through large openings (windows, doors) or cracks and gaps through building envelope and interzonal transport from other rooms cannot be neglected.

Settled particulate matter sampling was carried out by passive methods during 28 days. The adjusted sampling method for ambient air was used for indoor environment. The aerosol particulates were captured into Petri dishes (8.5 cm diameter), installed at three height levels: on the floor, at height of 0.8 m from the floor and at height of 2.2 m from the floor. The settling of particles proceeded onto both by water filled Petri dishes (wet gravitational settling) and empty Petri dishes (dry gravitational settling) at each monitored level. The particle total mass was calculated by gravimetric method from the Petri dish mass increases; the surface particle concentrations were determined by standard way.

Suspended particulate matter investigation was focused on total suspended particles (TSP) and thoracic fraction PM_{10}. Investigation was carried out in the same rooms in the investigated flat building in the city of Košice. Measurement have included integral particles sampling onto a collection material (membrane filter Synpor 0.83 μm pore size, 35 mm in diameter and PTFE filter for TSP and PM_{10}, respectively) by sampling equipment VPS 2000 (Envitech, Trenčín) at the constant air flow of 600 litres/hour during a sampling period of approximately 24 hours. Because of minimization of humidity interference and volatile organic matters elimination, the filters were dried at a temperature of 105°C for 8 h before sampling than equilibrated at a constant temperature and humidity (e.g. 20°C and 50% RH) for 24 h before and after sampling. The particulate mass concentrations were determined by gravimetric method from the increase of filter weight (measured by analytical balance fy Mettler Toledo within 0.00001 g). The average concentrations of measured particulate matter in studied rooms are presented in Table1.

	Mean
Settled particulate matter - surface concentration [μg.cm^{-2}]	44.8
Total suspended particulates - mass concentration [μg.m^{-3}]	84.7
PM_{10} - mass concentration [μg.m^{-3}]	45.4
PM_{10} / TSP ratio	0.5

Table 1. The mean concentrations of settled and suspended particulate matter

The surface concentrations of settled particulate matter measured in selected rooms were in the range 7.0 to 86.6 μg.cm^{-2} while the average surface concentrations for the rooms were calculated from 32.7 to 63.9 μg.cm^{-2} (Table 2). The percentage of non-dissolved portion of settled particulate matter was calculated by dividing of the non-dissolved mass separated by filters by total deposited mass [47].

Room	Total deposited mass [μg]	Average surface concentration [μg.cm^{-2}]	Non-dissolved mass [μg]	Percentage of non-dissolved [%]
Kitchen	44.8 x 10^3	63.9	17.06 x 10^3	38.1
Living room	27.6 x 10^3	37.8	19.70 x 10^3	71.4
Working room	21.7 x 10^3	32.7	7.36 x 10^3	33.9

Table 2. Settled particulate matter and percentage of non-dissolved particles in total deposited mass

The highest total deposited mass was detected in the kitchen, the lowest in the working room (Table 2). The highest non-dissolved mass was expected as well. However, there was detected the highest percentage of non dissolved particulate matter in the living room. Fibres from carpets, textile and upholstered furniture represented the essential part of non-dissolved from the total deposited mass (Figure 1).

Figure 1. Non-dissolved particles captured on the filter

The results of indoor particle deposition monitoring considering the three high levels in all monitored rooms are summarized in Table 3. Besides the standard wet deposition, the dry deposition was included in the study in order to investigate the re-suspension processes. The surface concentrations of particles ranged from 21.0 to 86.6 µg.cm^{-2} by wet gravitational settling and from 7.0 to 39.5 µg.cm^{-2} by dry gravitational settling in all monitored rooms.

Surface concentration [µg.cm^{-2}]	Distance from the floor		
	0.0 m	0.8 m	2.2 m
Kitchen			
wet gravitational settling	86.62	53.50	51.59
dry gravitational settling	39.49	27.39	24.84
Living room			
wet gravitational settling	42.68	38.22	32.48
dry gravitational settling	27.39	21.02	14.01
Working room			
wet gravitational settling	47.77	29.29	21.02
dry gravitational settling	17.19	15.92	7.01

Table 3. Surface concentration of particulate matter

The highest surface concentrations of particulate matters were measured in the kitchen at all monitored levels. The surface concentration values were expected to be the highest in the kitchen because of the most intensive indoor particulate sources. The surface concentrations determined in the other rooms reached the comparable values.

The particles surface concentration was found to be decreased with the height of the room from the floor to the ceiling construction at wet gravitational settling in all monitored rooms

(Figure 2), as well as at dry gravitational settling (Figure 3). That means the lowest surface concentrations of particulates were measured at the height level of 2.2 m in all monitored rooms.

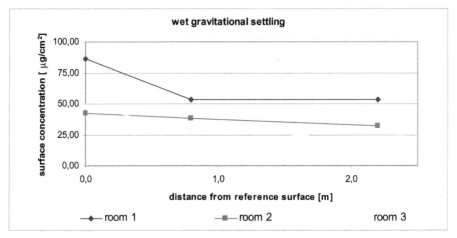

Figure 2. Particles surface concentration versus height level at wet gravitational settling

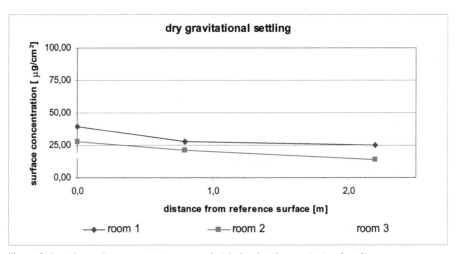

Figure 3. Particles surface concentration versus height level at dry gravitational settling

Particles re-suspension effect was studied in real conditions without boundary conditions providing for any effect elimination. The particles release was expressed in percentage; the amount of particulates settled into water filled Petri dishes was represented by 100%.

The proportion of particles (re-suspended) released into the air after sedimentation settling was calculated as a difference between surface concentrations at both wet and dry settling for each height level and all monitored rooms [48]. The particles portions in relation to the height level in monitored rooms are illustrated in Figure 4.

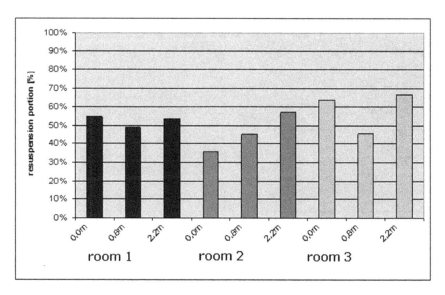

Figure 4. The particles portions in relation to the height level

The values of re-suspension particles portions ranged from 45.6 to 58.7% in monitored rooms. The results of particles re-suspension effect were not consistent with our expectations. None trend of particles release in relation to the height level was confirmed (Figure 4). The wide differences in particle re-suspension portions were achieved at monitored height levels in studied rooms: from 35.8 to 64 % on the floor and from 56.8 to 66.7 % at the height level of 2.2 m from the floor. The comparable portions for particles release was achieved only at the height level of 0.8 m from the floor (48.8, 45.0 and 45.7 %).The average values of re-suspended particles portion in all monitored rooms are presented in Table 4.

Room	Re-suspended portion [%]
Kitchen	51.69
Living room	45.89
Working room	58.77

Table 4. The re-suspended portions of particulate matter in monitored rooms

The non-expected conclusion has resulted from comparison of the average values of resuspension portions in monitored rooms. The highest portions of released particles were found out in the working room with a minimum operating mode (minimum people activity).

The mass concentrations of total suspended particulate matter (TSP) in studied rooms were detected in the range 59.028 to 114.583 µg.m^{-3}; PM$_{10}$ mass concentrations measured ranged from 31.94 to 55.56 µg.m^3 (Table 5). Unlike settled particulate matter monitoring, the highest concentration of total suspended particles as well as PM$_{10}$ fraction were measured in the living room.

Room	TSP [µg.m^{-3}]	PM$_{10}$ [µg.m^{-3}]	PM$_{10}$/TSP
Kitchen	80.556	48.611	0.60
Living room	114.583	55.556	0.48
Working room	59.028	31.944	0.54

Table 5. Suspended particulate matter concentration

The PM$_{10}$ hygienic limit (50 µg.m^{-3}) for indoor air in the Slovak Republic was exceeded in one measured room; the mean mass concentration detected was close to the limit. PM$_{10}$ concentration values reached about half of TSP concentration values (PM$_{10}$/TSP ratio 0.48 for the living room, 0.60 for the kitchen and 0.54 for the working room).

The similar mean concentration value of 63.3 µg m^{-3} monitored in 34 homes in Hong Kong has been reported in [25]. The lower indoor PM$_{10}$ concentration levels were measured in Athens (mean values for all residences was 35.0 ± 10.7 µg.m^{-3} during the warm period and 31.8 ± 7.8 µg.m^{-3} during the cold period), presenting no exceedance above the 50 µg.m^{-3} limit value [26]; whereas the authors in the study [27] referred much higher mean concentrations of 202 and 215 µg.m^{-3} in poor Bangladeshi households. The very high PM$_{10}$ levels were caused by using wood, dung and other biomass fuels for cooking.

4. The morphology of settled and suspended particulate matter

The morphology of settled as well as suspended particulate matter was investigated by electron scanning microscopy (SEM) with equipment Jeol JSM-35CF (Japan) at various extensions ranging from 90 to 5500. The scanning electron microscopy (SEM) micrographs represent the morphology of selected particles. As shown in Figures 5 to 9, the particles of irregular shapes and various sizes were observed in the sample of settled particulate matter.

Figure 5. Settled particulate matter morphology

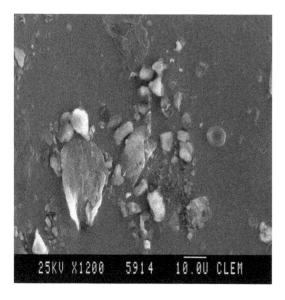

Figure 6. Settled particulate matter morphology

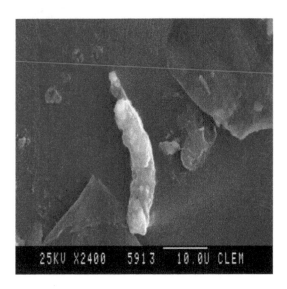

Figure 7. Detail of various shapes of settled particulate matter

Figure 8. Detail of various shapes of settled particulate matter

The majority of particles are non-spherical in shape with strong division of the surface. The occurrence of spherical as well as fibrous particles was not obvious.

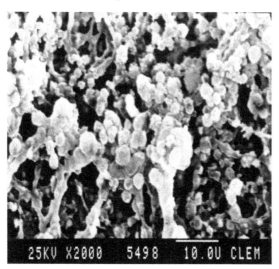

Figure 9. PM$_{10}$ particulate matter morphology

Individual particles along with the aggregates of fine particles were observed in PM$_{10}$ suspended particulate matter (Figure 9). The evaluation of SEM micrographs of the total suspended particulate samples showed that 80 - 90 % of the particles are smaller than 10 μm. In case of some samples, the particle size distribution was even shifted in the range of particle size under 5 μm. As referred by authors in the Chinese study [24], the analysis of the settled dusts collected in typical resident buildings showed that the volume percent for the fine particles (particle size < 10.5 μm) of the settled dusts ranged from 26 % - 38 %.

Seasonal variations and variations due to location were observed in both the morphological measurements and chemical analysis of settled dust collected inside the main foyers of three University buildings in Wolverhampton City Centre, U.K. [28].

5. The chemical composition of settled and suspended particulate matter

The elemental EDX analyses were carried out on the micro-analytical system LINK AN 10 000 operating in secondary mode at a potential 25 kV. The energy-dispersion X-ray system provided preliminary information on the elemental composition of the samples. The EDX spectra were very similar for majority of collected particulate matter samples. Principal inorganic elements constituting the particles calcium, silicon, aluminium, potassium, iron, chlorine, magnesium as well as titan and manganese were confirmed. The EDX spectrum in Figure 10 represents the elemental chemical composition of the settled particulate matter sample.

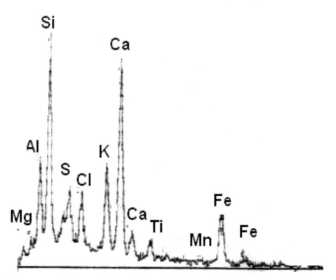

Figure 10. EDX spectrum of elemental chemical analysis of settled particulate matter

The energy-dispersive X-ray system interfaced to the SEM provides preliminary information on the elemental composition of the samples. Figure 11 presents the EDX spectrum of the suspended particulate matter sample.

In all samples discussed here, the EDX spectra were very similar for majority of collected particulate samples. The principal inorganic elements constituting the particles in order of peak intensity decreasing were Ca $\approx$ Si > O > Al > C > Mg > Fe > Cl > Na $\approx$ K. The presence of both carbon and oxygen, which can originate from organic compounds as well as from inorganic oxides, acids and/or salts, was confirmed [29].

The elements observed by EDX were confirmed also by using X- ray fluorescence analysis (XRF). The total amount of inorganic elements (except for carbon, oxygen and other elements with proton number under 11) in settled particulate matter measured by XRF was found very low and was about 2.23 %. In [30] organic carbon and elemental carbon made up 29 % and 2.5 % of the particulate matter, respectively. Water-soluble total carbon content in PM_{10} corresponds to 16% of the total particle masses measured in India. Organic matter is by far the major PM_{10} component besides mineral oxides. As observed in [31] major individual organic compounds quantified included series of alkanes, n-alkanoic acids, n-alkanals, alkan-2-ones and PAHs. Alkanes and ketones make up a significant fraction of particle-phase organic compounds, ranging from C_{11} to C_{26}, and C_9 to C_{19}, respectively. In addition, other organic compound classes have been identified, such as alkanols, esters, furans, lactones, amides, and nitriles [28]. The measured percentage content of measured elements is summarised in Table 6.

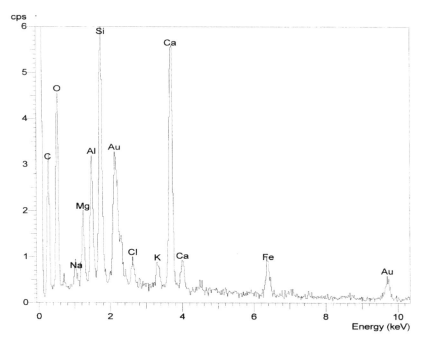

Figure 11. EDX spectrum of elemental chemical analysis of suspended particulate matter

Element	Minimum [%]	Maximum [%]	Mean [%]
Aluminium	0.14	0.18	0.159
Silicon	0.33	0.37	0.350
Phosphorous	0.02	0.03	0.021
Sulphur	0.30	0.31	0.306
Chlorine	0.48	0.66	0.570
Potassium	0.25	0.37	0.310
Calcium	0.33	0.51	0.420
Titane	0.02	0.03	0.025
Cromium	0.01	0.01	0.010
Manganese	0.004	0.006	0.005
Iron	0.01	0.03	0.020
Zinc	0.01	0.01	0.01
Bromium	0.002	0.0007	0.0014

Table 6. The percentage of basic inorganic elements measured by XRF in settled particulate matter

Chlorine, calcium, silicon, potassium and sulphur were found to be dominated; the concentrations of the other elements were quite lower as resulted from the quantitative XRF analysis (Table 6). The results of qualitative analysis by XRF correlated with those reported in [32]. The percentage of calcium and chromium measured by XRF is consistent with that measured by AAS (Table 9): 0.42 versus 0.43 % in case of calcium; 0.01 % by both XRF and AAS analysis in case of chromium. The XRF measured concentrations of iron and zinc were detected to be much lower than those detected by AAS (Table 9).

The principal component analysis shows the existents of three associations of the elements in settling particles: a) lithogenic (As, Co, Cr, Fe, lantanides and Sc); b) biogenic (Sr and Ca); c) authigenic (U and Se). The average element enrichment factors were higher in the first period of settled particulate matter sampling from: Se (739)> Zn (523)> Cr(105)> Br(104)> Sb(97)> As (69) [33]. The As, Br, Cr, Sb, Se, Sr and U average concentrations in the settled particulate matter were measured higher than their average crustal abundances [33].

Qualitative estimation of various functional groups in particulate matter proceeded with Fourier transformed infrared analysis FTIR (Figure 12).

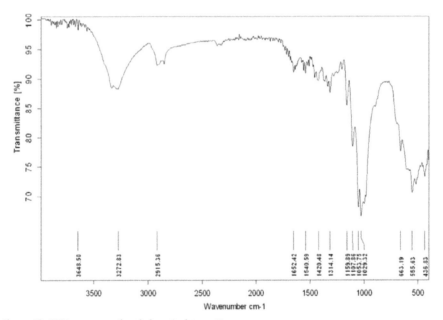

Figure 12. FTIR spectrum of settled particulate matter

Transmittances associated with particulate sulphate (near 618 and 1110 cm⁻¹), ammonium (2900–3200, 1430 cm⁻¹), hydroxyl (3200–3500 cm⁻¹), aliphatic carbon (2920 and 2850 cm⁻¹) and carbonyl (1650–1800 cm⁻¹) functional groups were observed. FTIR also identified several organic functional groups, although specific organic molecules could not be identified. In addition, there was also noticed the presence of inorganic nitrate (835 cm⁻¹) in [34].

Absorbances associated with sulphate, nitrate, ammonium, aliphatic carbon-hydrogen, and carbonyl functional groups as main constituent of particulate matter were observed also in the FTIR spectra of diesel generated PM_{10} [35]. The mass concentrations of sulphate, nitrate, ammonium, organic carbon (OC), elemental carbon (EC) were primarily measured in [36] in small particulate matter of size 0.1–3.0 μm.

The sum of Cl^-, NO_3^- and SO_4^{2-} concentrations represents a contribution of approximately 24% to the total mass in ambient PM_{10} as noticed in [37]. Compared to outdoors, indoor PM contained more silicate (36% of particle number), organic (29%, probably originating from human skin), and Ca-carbonate particles (12%) [38]. Indoor PM_{10} was elevated, chemically different and toxicologically more active than outdoor PM_{10} [38]. Suspended and settled particulate matter sampled in the child's bedroom was investigated in terms of mouse allergen in [39]. Airborne mouse allergen was detected in 48 of 57 (84%) bedrooms, and the median airborne mouse allergen concentration was 0.03 ng.m^{-3}. The median PM10 concentration was 48 mg.m^{-3} [39].

6. Metals content in suspended and settled particulate matter

The presence of selected metals in particulate matter samples was detected by atomic absorption spectrometry (SpectrAA-30, Varian, Austrália). Fe, Zn, and Cu were detected by a standard process in acetylene – air flame, Cd, Cr, Ni, Pb and Co were detected in graphite cell in the GTA 96 add-on equipment. Arsenic content was detected by hydride method in the VGA 76 add-on equipment.

Metals content was investigated in both settled and suspended particulate matter samples. Because of low quantity in the suspended particles samples, the metals concentrations were detected for TSP and PM_{10} filters all at once.

The results of AAS analysis of selected metals content in settled and suspended aerosols for each monitored room are presented as metal concentrations in Tables 7 and 8. The average concentrations of metals measured in insufficient amount for individual concentration detection for each room are presented for arsenic, cadmium, chromium, nickel and lead.

Metal	Kitchen	Living room	Working room
Calcium	0.64	1.46	2.06
Copper	0.04	0.06	0.07
Iron	2.56	1.78	4.73
Magnesium	0.22	0.50	0.67
Zinc	0.29	2.80	0.91
	Average concentration		
Arsenic	0.10		
Cadmium	0.03		
Chromium	0.04		
Nickel	0.05		
Lead	0.09		

Table 7. Surface metal concentrations in settled particulate matter [μg.cm^{-2}]

Metal	Kitchen	Living room	Working room
Calcium	2.16	2.47	1.56
Copper	0.18	0.10	0.16
Iron	0.14	0.51	1.71
Magnesium	0.41	0.51	0.37
Zinc	0.20	0.21	0.20
	Average concentration		
Arsenic	0.28		
Cadmium	0.07		
Chromium	0.10		
Nickel	0.14		
Lead	0.24		

Table 8. Mass metal concentrations in suspended particulate matter [μg.m^{-3}]

The surface metal concentrations of settled particulate matter were detected in the range from 0.03 (cadmium) to 4.73 μg.cm^{-2} (iron). The high concentrations were measured also in case of calcium and zinc. The highest concentrations were measured in case of iron, calcium and zinc. There were no significant differences of metal surface concentrations found out in all measured rooms. The metal concentration of the other investigated metals (Cr, Ni, Pb, Cd, As) in settled particulate matter were close to the detection limit (Table 7). The significant high concentrations of cadmium, chromium, arsenic and lead as tobacco smoke emissions were not confirmed in settled particulate matter.

The mass metal concentrations in suspended particulate matter range from 0.07 (cadmium) to 2.47 μg.m^{-3} (calcium). Similarly to settling PM metal concentrations, no significant differences were measured for the monitored rooms.

The percentage of studied metals content was calculated in settled as well as suspended particulate matter as the ratio of measured metal concentration to the particulate matter concentration (Table 9).

Metal	Settled PM [%]	Suspended PM [%]	Suspended/ settled metals
Arsenic	0.03	0.46	15.3
Cadmium	0.01	0.11	11.0
Chromium	0.01	0.17	17.0
Nickel	0.02	0.23	11.5
Lead	0.03	0.40	13.3
Calcium	0.43	3.25	7.6
Copper	0.02	0.26	13.0
Iron	0.88	0.33	0.4
Magnesium	0.14	0.68	4.9
Zinc	1.50	0.34	0.2

Table 9. The metals percentage content in settled and suspended particulate matter

The higher percentage of metals content was detected in suspended particulate matter in comparison to the settled particles. This finding may result from the fact that most of metals are cumulated in the finest fraction of aerosols [40] represented by suspended PM_{10} in this study. As reported in [41] Na, Al, Ca, Fe, Mg, Mn and Ti were found in coarse particles, while K, V, Cr, Ni, Cu, Zn, Cd, Sn, Pb, As and Se occured more in fine particles. In reference [44] there is noted that elements mostly concentrated in coarse mode including Al, Mg, Ca, Sc, Ti, Fe, Sr, Zr and Ba; elements mostly concentrated in accumulation mode including S, As, Se, Ag, Cd, Tl and Pb; and the elements having muti-mode distribution including Be, Na, K, Cr, Mn, Co, Ni, Cu, Zn, Ga, Mo, Sn and Sb.

The measured values of metals content in suspended particulate matter were 4.9 – 15.3 times higher for all metals except for iron and zinc. The comparison of the percentage content of arsenic, cadmium, chromium, nickel and lead in settled and suspended indoor particulate matter is presented in Figure 13.

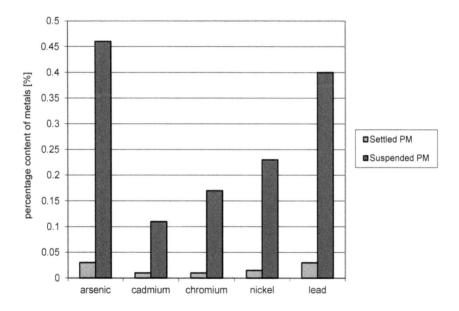

Figure 13. The percentage content of metals in settled and suspended particulate matter

The measured mass of metals contents in the samples of settled as well as suspended particulate were compared to the total mass of monitored particulate matter for each monitored room. Figures 14 and 15 represent the percentage content of metals in settled and suspended particulate matter for each monitored room, respectively.

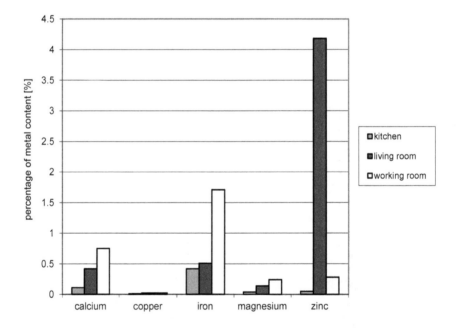

Figure 14. The percentage content of metals in settled particulate matter for monitored rooms

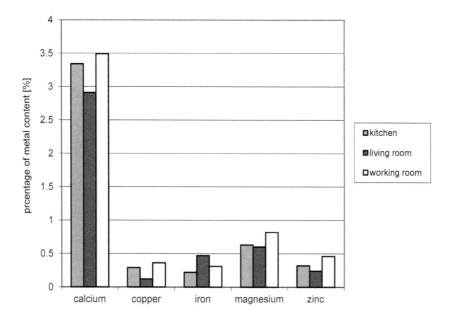

Figure 15. The percentage content of metals in suspended particulate matter for monitored rooms

The obtained mass concentrations of metals in indoor particle samples correspond with those in the typical urban aerosol [42,43]. The average indoor concentrations of total elements were lower than or comparable to those measured outdoors, suggesting that indoor elements originated mainly from outdoor emission sources. On the contrary, the authors in [24] reported the metal elements concentrations analyzed were 3-15 times higher relative to soil background values in China.

Anthropogenic sources include fossil fuel combustion, industrial metallurgical processes, vehicle emission and waste incinerations. Natural sources include a variety of processes acting on crustal minerals, such as volcanism, erosion and surface winds, as well as from forest fires and the oceans. Some elements are potentially toxic trace metals, such as Pb, Cd, V, Fe, Zn, Cr, Ni, Mn and Cu.

7. Conclusion

Particulate matter exposure that occurs indoors probably constitutes a significant fraction of the overall exposure to hazardous particles since typically people spend most time indoors. The indoor settled as well as suspended particulate matter was monitored and particles morphology and chemical composition with special regard to metal content investigation was performed.

- Particles of irregular shapes and various sizes were observed in settled as well as suspended particulate matter.
- Principal inorganic elements constituting the particulate matter such as calcium, silicon, aluminium, potassium, iron, chlorine, magnesium as well as titan and manganese was confirmed. The percentage of inorganic elements mentioned was detected very low in the range of 2.23 %.
- Higher percentage of metals content was detected in suspended particulate matter in comparison to the settled particles except for iron and zinc. The measured values of metals content were 4.9 – 15.3 times higher in suspended particulate matter when comparing to the settled one.
- There were found out no significant differences of metal surface concentrations in the measured rooms in spite of the various indoor particulate matter sources.

The results demonstrate the complexity of indoor particulate matter nature affecting their surface properties. The results also emphasise the need for further research to a more complete understanding of the chemical nature of indoor particulate matter in connection with their surface reactivity. Due to the negative biological influence of particulate matters and their specific properties resulting in synergic effect of the other pollutants in the indoor air it is necessary to investigate the ways of indoor particulate matters occurrence minimization and/or elimination.

Author details

Adriana Estokova* and Nadezda Stevulova
*Technical University of Kosice, Faculty of Civil Engineering,
Institute of Environmental Engineering, Kosice, Slovakia*

* Corresponding Author

Acknowledgement

This research has been carried out in terms of the project NFP 26220120037 and NFP 26220120018 supported from the European Union Structural funds. The authors would like to thank following people for their help and suggestions: assoc. prof. Dr. Magdalena Balintova, Ing. Lenka Kubincova, PhD. and Dr. Gabriel Janak from Technical University of Kosice, Slovakia.

8. References

[1]   Hinds WC (1999) Aerosol technology: Properties, Behavior and Measurement of Airborne Particles. New York: Wiley. 483 p.

[2]   Morawska L, Salthammer T (2003) Indoor Environment: Airborne Particles and Settled Dust. Weinheim: Wiley-VCH. 450 p.

[3]   Seinfeld J.H (1986) Atmospheric Chemistry and Physics of Air Pollution. New York: Wiley. 738 p.

[4]   Klinda J, Lieskovská Z (2009) Report on State of the Environment of the Slovak Republic in 2009. Bratislava: SAZP. 280 p.

[5]   McCormack MC, Breysse PN, Matsui EC (2011) Indoor Particulate Matter Increases Asthma Morbidity in Children with Non-Atopic and Atopic Asthma. Ann. Allergy Asthma Immunol. 106: 308–315.

[6]   Pope CA, Dockery DW (2006) Health Effects of Fine Particulate Air Pollution: Lines that Connect. J. Air Waste Manage. Assoc. 56: 709–742.

[7]   Linares C, Diaz J (2010) Short-term Effect of Concentrations of Fine Particulate Matter on Hospital Admissions Due to Cardiovascular and Respiratory Causes among the over-75 age Group in Madrid, Spain. Public Health. 124: 28 – 36.

[8]   Faustini A, Stafoggia M, Berti G, Bisanti L, Chiusolo M, Cernigliaro A, Mallone S, Primerano R, Scarnato C, Simonato L, Vigotti MA, Forastiere F (2011) The Relationship Between Ambient Particulate Matter and Respiratory Mortality: a Multi-city Study in Italy. Eur. Respir. J. 38: 538-47.

[9]   Wallace L (1996) Indoor Particles: a Review. J. Air and Waste Manag. Assoc. 42: 98-126.

[10] Sousa S, Alvim Ferraz MCM, Martins FG (2012) Indoor PM10 and PM2.5 at Nurseries and Primary Schools. Advanced Materials Research. 385: 433-440.

[11] Branis M, Safranek J, Hytychova A (2011) Indoor and Outdoor Sources of Size-Resolved Mass Concentration of Particulate Matter in a School Gym-Implication for Exposure of Exercising Children. Environ. Sci. Pollut. Res. Int. 18: 598-609.

[12] Estokova A, Stevulova N, Kubincova L (2008) Indoor aerosol examining. J. Chemické listy. 102: 361-362.

[13] Klepeis NE, Nelson WC, Ott WR, Robinson JP, Tsang AM (2001) The National Human Activity Pattern Survey (NHAPS): a Resource for Assessing Exposure to Environmental Pollutants. Journal of Exposure Analysis and Environmental Epidemiology. 11: 231–252.

[14] Diapouli E, Chaloulakou A, Spyrellis N (2008) Indoor and Outdoor PM Concentrations at a Residential Environment, in the Athens Area. Global NEST j. 10: pp 201 – 208.

[15] Thatcher TL, Lunden MM, Revzan KL, Sextro RG, Brown NJ (2003) A Concentration Rebound Method for Measuring Particle Penetration and Deposition in the Indoor. Aerosol Sci. Techn. 37: 847 – 864.

[16] Raaschou-Nielsen O, Sorensen M, Hertel O, Chawes BK, Vissing N, Bonnelykke K, Bisgaard H (2011) Predictors of Indoor Fine Particulate Matter in Infants' Bedrooms in Denmar. Environ. Res. 111: 87–93.

[17] Mueller D, Semple S, Garden C, Coggins M, Galea KS, Whelan P, Cowie H, Sánchez-Jiménez A, Thorne PS, Hurley JF, Ayres JG (2011) Contribution of Solid Fuel, Gas Combustion, or Tobacco Smoke to Indoor Air Pollutant Concentrations in Irish and Scottish Homes. Indoor Air, DOI: 10.1111/j.1600-0668.2011.00755.x.

[18] Mueller D, Uibel S, Braun M, Klingelhoefer D, Takemura M, Groneberg DA (2011) Tobacco Smoke Particles and Indoor Air Quality (ToPIQ) - the Protocol of a New Study. J. Occup. Med. Toxicol. 6: 35.

[19] Thatcher TL, McKone TE, Fisk WJ, Sohn MD, Delp WW, Riley WJ, Sextro RG (2001) Factors Affecting the Concentration of Outdoor Particles Indoors (COPI): Identification of Data Needs and Existing Data. Report Number LBNL—49321.

[20] Thatcher TL, Lai CK, Moreno JR, Sextro RG, Nazaroff WW (2001) Effects of room furnishings and air speed on particle deposition rates indoors. Lawrence Berkeley National Laboratory, Report LBNL-48414, Berkeley USA, 2001.

[21] Thatcher TL, Layton DW (1995) Deposition, resuspension, and penetration, of particles within a residence. J. Atmospheric Environment. 29: 1487-1497.

[22] Crump JG, Flagan RC, Seinfeld JH (1983) Particle wall loss rates in vessels. Aerosol Sci. Technol. 2: 303–309.

[23] Zhao B, Wu J (2007) Particle Deposition in Indoor Environments: Analysis of Influencing Factors. J. of Hazardous Materials. 47: 439–448.

[24] Zeng L, Cao Z, Yue J, Lu J, Zhang C (2011) Depth Profiles of Particulate Matter and Elements in Particulate Matter in Xi'an, China. (iCBBE) 2011 5th International Conference on Bioinformatics and Biomedical Engineering, Wuhan 10-12 May 2011, p1-4.

[25] Chao CY, Wong KK (2002) Residential Indoor PM_{10} and $PM_{2.5}$ in Hong Kong and the Elemental Composition. Atmospheric Environ. 36: 265–277.

[26] Diapouli E, Eleftheriadis K, Karanasiou A, Vratolis S, Hermansen O, Colbeck I, Lazaridis M (2011) Indoor and Outdoor Particle Number and Mass Concentrations in Athens. Sources, Sinks and Variability of Aerosol Parameters. Aerosol and Air Quality Research, 11: 632–642.

[27] Dasgupta S, Huq M, Khaliquzzaman M, Pandey K, Wheeler D (2004) Indoor Air Quality for Poor Families: New Evidence from Bangladesh. Development Research Group, World Bank

[28] Shilton V, Giess P, Mitchell D, Williams C (2002) The Characterisation of Settled Dust by Scanning Electron Microscopy and Energy Dispersive X-ray Analysis . Water, Air, & Soil Pollution. 2: 237-246.

[29] Estokova A, Stevulova N, Kubincova L (2010) Particulate Matter Investigation in Indoor Environment. Global NEST J. 12: 20-26.

[30] Fermo P, Piazzalunga A, Vecchi R, Valli G, Ceriani M (2006) A TGA/FT-IR Study for Measuring OC and EC in Aerosol Samples. Atmos. Chem. Phys. 6: 255-266.

[31] He LY, Hu M, Wang L, Huangn XF, Zhangn YH (2004) Characterization of Fine Organic Particulate Matter from Chinese Cooking. J. Environmental Sciences China. 16: 570-575.

[32] López ML, Ceppi S, Palancar GG, Olcese LE, Tirao G, Toselli BM (2011) Elemental Concentration and Source Identification of PM10 and PM2.5 by SR-XRF in Córdoba City, Argentina. Atmos. Environ. 45: 5450-5457.

[33] Leal Acosta ML, Choumiline E, Silverberg N (2009) Concentrations of Trace Elements in Settling Particulate Matter and Particulate Element Fluxes in the Concepción Bay, the Gulf of California during the years 1996-1998. American Geophysical Union, Fall Meeting 2009, abstract #EP21A-0581.

[34] Tiwary A, Reff A, Colls JJ (2008) Collection of Ambient Particulate Matter by Porous Vegetation Barriers: Sampling and Characterization Methods. Aerosol Sci. 39: 40– 47.

[35] Kinsey J, Mitchell W, Squier W, Linna K, King F (2006) Evaluation of Methods for the Determination of Diesel-generated Fine Particulate Matter: Physical Characterization Results. J. Aerosol Sci. 37: 63-87.

[36] Jeong CH, McGuire ML, Godri KJ, Slowik JG, Rehbein PJG, Evans GJ (2011) Quantification of Aerosol Chemical Composition Using Continuous Single Particle Measurements. Atmos. Chem. Phys. 11: 7027–7044.

[37] Bogo H, Otero M, Castro P, Ozafrán MJ, Kreiner A, Calvo EJ, Neg RM (2003) Study of Atmospheric Particulate Matter in Buenos Aires City. Atmos. Environ. 37: 1135–1147.

[38] Oeder S, Dietrich S, Weichenmeier I, Schober W, Pusch G, Jörres RA, Schierl R, Nowak D, Fromme H, Behrendt H, Buters JT (2011) Toxicity and Elemental Composition of Particulate Matter from Outdoor and Indoor Air of Elementary Schools in Munich, Germany. Indoor Air. 9, DOI: 10.1111/j.1600-0668.2011.00743.x.

[39] Matsui EC, Simons E, Rand C, Butz A, Buckley TJ, Breysse P, Eggleston PA (2005) Airborne Mouse Allergen in the Homes of Inner-City Children with Asthma. J. Allergy and Clinical Immunology. 115: 358-363.

[40] Asmi A et al. (2000) Connection Between Ultra-Fine Aerosols Indoors and Outdoors in an Office Environment. In: Proc of Conference Healthy Buildings 2000, Helsinky, pp. 543 - 548.

[41] Wang X, Bi X, Sheng G, Fu J (2006) Hospital indoor PM10/PM2.5 and Associated Trace Elements in Guangzhou, China. Sci Total Environ. 366(1): 124-35.

[42] Hanculak J, Bobro M, Slanco P, Brehuv J (2005) Heavy Metals in Solid Immissions in the Area of Jelšava. Chem.Listy. 99: 141-142.

[43] Takac P, Szabova T, Kozakova L, Benkova M (2009) Heavy Metals and their Bioavailability from Soils in the Long-term Polluted Central Spiš Region of SR. Plant, Soil and Environment. 55: 167-172.

[44] Duan J, Tan J, Wang S, Hao J, Chai F (2012) Size Distributions and Sources of Elements in Particulate Matter at Curbside, Urban and Rural Sites in Beijing. J. Environmental Sciences. 24(1): 87–94.

[45] Freitas MC, Canha N, Martinho M, Almeida-Silva M, Almeida SM, Pegas P, Alves C, Pio C, Trancoso M, Sousa R, Mouro F, Contreiras T (2011) Indoor air quality in primary schools. In: Moldoveanu AM, editor. Advanced Topics in Environmental Health and Air Pollution Case Studies. Rijeka: InTech. pp. 361-384.

[46] Hunt GT (2011) Nuisance dusts - validation and application of a novel dry deposition method for total dust fall. In: Mazzeo NA, editor. Air Quality Monitoring, Assessment and Management. Rijeka: InTech. pp. 77-92.

[47] Kubincova L, Estokova A, Stevulova N (2008) Metal substances content in the indoor particulate matter - the case study. Selected Scientific Papers. 3(2): 33-40.

[48] Kubincova L, Estokova A, Stevulova N (2008) Aerosols deposition onto horizontally oriented indoor surfaces - the case study. Selected Scientific Papers. Roč. 3(1): 89-96.

Permissions

The contributors of this book come from diverse backgrounds, making this book a truly international effort. This book will bring forth new frontiers with its revolutionizing research information and detailed analysis of the nascent developments around the world.

We would like to thank Hayder Abdul-Razzak, for lending his expertise to make the book truly unique. He has played a crucial role in the development of this book. Without his invaluable contribution this book wouldn't have been possible. He has made vital efforts to compile up to date information on the varied aspects of this subject to make this book a valuable addition to the collection of many professionals and students.

This book was conceptualized with the vision of imparting up-to-date information and advanced data in this field. To ensure the same, a matchless editorial board was set up. Every individual on the board went through rigorous rounds of assessment to prove their worth. After which they invested a large part of their time researching and compiling the most relevant data for our readers. Conferences and sessions were held from time to time between the editorial board and the contributing authors to present the data in the most comprehensible form. The editorial team has worked tirelessly to provide valuable and valid information to help people across the globe.

Every chapter published in this book has been scrutinized by our experts. Their significance has been extensively debated. The topics covered herein carry significant findings which will fuel the growth of the discipline. They may even be implemented as practical applications or may be referred to as a beginning point for another development. Chapters in this book were first published by InTech; hereby published with permission under the Creative Commons Attribution License or equivalent.

The editorial board has been involved in producing this book since its inception. They have spent rigorous hours researching and exploring the diverse topics which have resulted in the successful publishing of this book. They have passed on their knowledge of decades through this book. To expedite this challenging task, the publisher supported the team at every step. A small team of assistant editors was also appointed to further simplify the editing procedure and attain best results for the readers.

Our editorial team has been hand-picked from every corner of the world. Their multi-ethnicity adds dynamic inputs to the discussions which result in innovative outcomes. These outcomes are then further discussed with the researchers and contributors who give their valuable feedback and opinion regarding the same. The feedback is then collaborated with the researches and they are edited in a comprehensive manner to aid the understanding of the subject.

Apart from the editorial board, the designing team has also invested a significant amount of their time in understanding the subject and creating the most relevant covers. They scrutinized every image to scout for the most suitable representation of the subject and create an appropriate cover for the book.

The publishing team has been involved in this book since its early stages. They were actively engaged in every process, be it collecting the data, connecting with the contributors or procuring relevant information. The team has been an ardent support to the editorial, designing and production team. Their endless efforts to recruit the best for this project, has resulted in the accomplishment of this book. They are a veteran in the field of academics and their pool of knowledge is as vast as their experience in printing. Their expertise and guidance has proved useful at every step. Their uncompromising quality standards have made this book an exceptional effort. Their encouragement from time to time has been an inspiration for everyone.

The publisher and the editorial board hope that this book will prove to be a valuable piece of knowledge for researchers, students, practitioners and scholars across the globe.

List of Contributors

Gerhard Held and Ana Maria Gomes
Instituto de Pesquisas Meteorológicas, Universidade Estadual Paulista, Bauru, S.P., Brazil

Andrew G. Allen and Arnaldo A. Cardoso
Instituto de Química, Universidade Estadual Paulista, Araraquara, S.P., Brazil

Fabio J.S. Lopes and Eduardo Landulfo
Centro de Lasers e Aplicações, Instituto de Pesquisas Energéticas e Nucleares, Universidade de São Paulo, São Paulo, Brazil

Mohd Zul Helmi Rozaini
School of Environmental Sciences, University of East Anglia, Norwich, Norfolk, UK
Department of Chemical Sciences, University Malaysia Terengganu, Kuala Terengganu, Terengganu, Malaysia

Biwu Chu, Jingkun Jiang, Zifeng Lu, Kun Wang, Junhua Li and Jiming Hao
State Key Laboratory of Environment Simulation and Pollution Control, School of Environment, Tsinghua University, Beijing, China

Chul Eddy Chung
Gwangju Institute of Science and Technology, Republic of Korea

Shexia Ma
Center for Research on Urban Environment, South China Institute of Environmental Sciences (SCIES), Ministry of Environmental Protection (MEP), Guangzhou, China

Gourihar Kulkarni
Pacific Northwest National Laboratory, Richland, WA, USA

Shiyong Shao, Yinbo Huang and Ruizhong Rao
Key Laboratory of Atmospheric Composition and Optical Radiation, Anhui Institute of Optics and Fine Mechanics, Chinese Academy of Sciences, China

Karel Klouda
State Office for Nuclear Safety, Czech Republic

Stanislav Brádka and Petr Otáhal
The National Institute for Nuclear, Chemical and Biological Protection, Kosice, Slovakia

Adriana Estokova and Nadezda Stevulova
Technical University of Kosice, Faculty of Civil Engineering, Institute of Environmental Engineering, Kosice, Slovakia

.

Printed in the USA
CPSIA information can be obtained
at www.ICGtesting.com
JSHW011409221024
72173JS00003B/476

9 781632 390820